Tōjisha Manga

"By identifying the genre *tōjisha manga* and providing the foundation for its analysis, Yoshiko Okuyama offers a valuable and necessary intervention into our understanding of manga, autobiographical comics, and disability. Written with a combination of care, empathy, and rigorous scholarship, *Tōjisha Manga: Japan's Graphic Memoirs of Brain and Mental Health* will resonate in comics studies and graphic medicine, exposing readers to fascinating, underappreciated works."

—Andrew J. Kunka, *Professor at the University of South Carolina Sumter, USA*

"Yoshiko Okuyama brings together insights gleaned from North American and Japanese disability studies, the Japanese disability rights movement, and interviews with manga artists to create an innovative and highly original study of what she calls '*tōjisha manga*'—graphic memoirs written by people living with various mental disabilities and their families. Presented in an accessible, conversational tone, *Tōjisha Manga* provides an exemplary, interdisciplinary foundation for future considerations of these important cultural artifacts and the women and men who produce them."

—Michael Rembis, *Associate Professor of History and the Director of the Center for Disability Studies at the University at Buffalo, USA*

Yoshiko Okuyama

Tōjisha Manga

Japan's Graphic Memoirs of Brain and Mental Health

palgrave
macmillan

Yoshiko Okuyama
Department of Languages
University of Hawai'i at Hilo
Hilo, HI, USA

ISBN 978-3-031-00839-9 ISBN 978-3-031-00840-5 (eBook)
https://doi.org/10.1007/978-3-031-00840-5

From Utsunuke: Utsu wo nuketa hitotachi by Tanaka Keiichi, published by KADOKAWA, 2017

This Palgrave Macmillan imprint is published by the registered company Springer Nature Switzerland AG.
The registered company address is: Gewerbestrasse 11, 6330 Cham, Switzerland

This book is dedicated to my mother, Okuyama Hatsue, whose life has taught me many precious lessons. This book would not have been written without her presence in my life.

Conventions

Japanese terms and other foreign words are italicized (e.g., *gaki* in Japanese) in the Hepburn style with macrons for long vowels (e.g., *tōjisha*) and are followed by the English translation in parentheses or vice versa. This practice excludes locational terms such as Tokyo and Osaka as well as Japanese words that commonly appear in English writing such as manga, anime, and otaku. Names of Japanese people such as manga authors are written using the Japanese convention of placing the family name first. For instance, the name of the manga artist Tezuka Osamu will appear in original Japanese order, not in the English format, Osamu Tezuka. The same convention is followed in every part of this book, including indexes, except for "Acknowledgments" in which the Western reading of Japanese names may be used along with non-Japanese names in English style for consistency.

Regarding usage of the controversial term *disability*, the definition of the Americans with Disabilities Act (ADA) has been applied in reference to a person with a disability: "a person who has a physical or mental impairment that substantially limits one or more major life activities, a person who has a history or record of such an impairment, or a person who is perceived by others as having such an impairment" (https://www.ada.gov/cguide.htm). However, unless they are used in direct quotes, terms such as *mental illness* and *mental disorder* are avoided for considerations articulated in "A Word About Language" in "Introduction" chapter. The author is well aware that those expressions have political implications and that, as some groups such as madness studies scholars and activists argue, conditions such as depression or Asperger's syndrome should not be labeled a disorder. Supporting and respecting those views, the author instead uses more nuanced phrases such as *mental health problem* or *neurological condition* unless the term is part of the clinical label of the condition that is more commonly known to the general readership (e.g., obsessive-compulsive disorder). This decision is intended to depathologize the very condition and avoid reinforcing ableists' perspectives. A more detailed account of language use can be found in "Introduction" chapter.

ACKNOWLEDGMENTS

Tōjisha Manga is the outcome of years of my research in the fields of disability studies and popular culture and the fruit of my personal and professional involvement in these areas of studies, particularly on the topic of mental disability. The publication of this book would not have been possible without the generous efforts and assistance of the many people and organizations who supported this book project.

First, I want to thank my family for their important presence in my life. My mother, Hatsue Okuyama, has been a source of inspiration for this book, particularly because of her lifelong struggle with disability. The experience of living with her under the influence of her psychological struggles had a great impact on the way that I, as both a scholar and a caretaker, view many of the mental health issues discussed in this book. On many occasions, I felt as if I had been stranded with no help in sight until Dr. Shelley Ham came to our rescue. I also thank my son, Sai O. Weiss, for his genuine and unfailing praise of my research and my husband, David Weiss, for his tireless love and support for this project despite the many occasions that my research and writing took me away from our time together.

Second, I would like to acknowledge key individuals in this book project. Cristina Bacchilega, retired professor of English at the University of Hawai'i at Mānoa, is the first person who urged me to explore a connection between popular culture and disability studies. The idea of investigating the representation of disability in manga occurred to me during our collegial conversation in Cristina's office one afternoon several years ago.

Cristina, this is the second book resulting from that casual chat! Another influential person is Michael Rembis, associate professor, Department of History, and director of the Center for Disability Studies, College of Arts and Sciences, University at Buffalo, State University of New York. I met Michael when he lectured on the history of mental disability at the National Endowment for the Humanities (NEH) 2018 Summer Institute on the Global Histories of Disabilities. Our post-lecture conversation on mad studies helped me to narrow my focus on the representation of mental health issues in manga for a new book project. Furthermore, Michael generously took his time to read my draft chapters of Part I and provided invaluable comments that led to significant improvement in the revision in fall 2021. In addition, I am forever indebted to my dear friend and proofreader, Catherine Collier, who read and edited numerous drafts of the manuscript from 2020 through 2022. Having lived experience of some of the conditions discussed in this book, she considered this book her priority in spite of the many responsibilities already on her shoulders. Cathy, thank you for reading my drafts intensively for hours, catching my grammar and factual errors, and also being a light guiding me through my challenging times of writing during the pandemic.

I also acknowledge many other scholars and friends who are also influential and supportive of me from the conception to the completion of this book project. Those precious souls include but are not limited to Linda Ambrosia, Pam Bond, Andrew Campana, Yukie Chihara, Mark Goodman, Kathy Hammes, Sonia Juvik, Christina Osterwalder, Hyunhee Park, and Yukari Kunisue. I am truly grateful for their positive force in my professional or personal life before and during the pandemic years. I particularly appreciate Andrew and Linda for reading some of the chapters in Part II and providing valuable feedback. It is also important to note that an earlier version of chapter "Nonami Tsuna's *Akira-san* Series (2011–2017): A Memoir of a "Cassandra" Wife" from this book appeared in *U.S.-Japan Women's Journal*, and the University of Hawai'i Press, which presides over the journal, provided permission for the reprint. I am grateful to editor-in-chief, Alisa Freeman, for her assistance in refining the chapter and obtaining the permission.

Third, equally important to this project have been my editors and assistant editors at Palgrave Macmillan for their unflappable faith in my ability to write and complete this book. Shaun Virgil, senior editor of Cultural, Media, and Communication Studies at Palgrave Macmillan, approached me by email prior to my presentation at the 2018 Association for Asian

Studies (AAS) conference in Washington, DC. After I discussed my book project with him during the conference that year, my partnership with Palgrave Macmillan officially began. Later, my book project was transferred into the capable hands of Camille "Milly" Davies, another editor in Cultural, Media, and Communication Studies. Due to research and personal issues brought by COVID-19, I had to ask her to extend my manuscript delivery due date multiple times. Milly's unfailing patience with me has been essential to the completion of this publication. I also thank her capable editorial assistant, Imogen Higgins, for providing me with consistent editorial support and excellent guidance through prompt email communication in spite of the substantial time difference between London and Hawai'i. Without you, Immy, we would not have reached the finish line! Similarly, I am grateful to Palgrave's external reviewers not only for their extremely useful and practical comments but also for their enthusiasm for this project.

Fourth, I'd like to acknowledge the important role played by several organizations in this book project. The project was financially supported by the University of Hawai'i at Hilo's Research Relations Grant award in 2018 as well as by a Japan Research Travel Grant award from the AAS's Northeast Asia Council (NEAC) in 2019. I was efficiently assisted in my archival research on manga by the staff at the Kyoto International Manga Museum located in Nakagyo-ku, Kyoto, Japan. I also benefited from the great opportunity to study the world history of mental disability at NEH's 2018 Summer Institute on the Global Histories of Disabilities and to participate in a panel where I presented a paper "Destigmatizing Mental Illness in *Tōjisha* Manga" at a 2019 AAS annual conference. Furthermore, Kyoto Seika University, a private university well-known for manga and anime studies, helped arrange my interviews with three professors of manga studies at their school. My most heartfelt thanks go to Gentosha, KADOKAWA, Kodansha, Komikku Shuppan, and Seiwa Shoten, the publishers of the manga titles analyzed in this book, for helping arrange my interviews with the featured manga artists and permitting me to use the artists' manga panels, including the one for the book's cover design.

Last but not least, I'd like to mention special thanks to the manga artists, Okita Bakka (chapter "Okita Bakka's *Gaki-Tame* Series (2011–2013): A Memoir of a "Troublemaker" Aspie Girl"), Nonami Tsuna (chapter "Nonami Tsuna's *Akira-san* Series (2011–2017): A Memoir of a "Cassandra" Wife"), Tanaka Keiichi (chapter "Tanaka Keiichi's *Utsunuke* (2017): An Ode to Depression Tōjisha"), Hosokawa Tenten (chapter

"Hosokawa Tenten's *Tsure utsu* Series (2006–2013): A Couple's Lived Experience of Depression"), and Shiramizu Sadako (chapter "Shiramizu Sadako's *Uchi no OCD* (2015): A Collaborative Memoir of Obsessive-Compulsive Disorder"), who agreed to be interviewed and have their drawings used for this book. Similarly, Darin Igawa, Graphics Coordinator of the University of Hawai'i at Hilo's Graphics Services, assisted me in producing high-quality graphic images for the selected manga titles at the final stage of my manuscript writing. Thanks to all who have assisted with, contributed to, and believed in the publication of this book until the very end.

Contents

1 Introduction 1

Part I Foundations 19

2 Tōjisha 21

3 Tōjisha Narratives 51

4 Essay Manga: Japan's Autobiographical Comics 77

Part II Case Studies 111

5 Okita Bakka's *Gaki-tame* Series (2011–2013): A Memoir of a "Troublemaker" Aspie Girl 113

6 Nonami Tsuna's *Akira-san* Series (2011–2017): A Memoir of a "Cassandra" Wife 145

7 Tanaka Keiichi's *Utsunuke* (2017): An Ode to Depression *Tōjisha* 181

8 Hosokawa Tenten's *Tsure utsu* Series (2006–2013): A Couple's Lived Experience of Depression 217

9 Shiramizu Sadako's *Uchi no OCD* (2015): A Collaborative Memoir of Obsessive-Compulsive Disorder 249

10 Afterword 279

Index 289

List of Figures

Fig. 1.1	Image by Hosokawa Tenten. (Courtesy of Gentosha)	7
Fig. 5.1	Image by Okita Bakka. (Courtesy of Kodansha)	114
Fig. 5.2	Image by Okita Bakka. (Courtesy of Kodansha)	126
Fig. 5.3	Image by Okita Bakka. (Courtesy of Kodansha)	128
Fig. 5.4	Image by Okita Bakka. (Courtesy of Kodansha)	131
Fig. 5.5	Image by Okita Bakka. (Courtesy of Kodansha)	132
Fig. 5.6	Image by Okita Bakka. (Courtesy of Kodansha)	134
Fig. 5.7	Image by Okita Bakka. (Courtesy of Kodansha)	135
Fig. 5.8	Image by Okita Bakka. (Courtesy of Kodansha)	138
Fig. 6.1	Image by Nonami Tsuna. (Courtesy of Kosumikku Shuppan)	146
Fig. 6.2	Image by Nonami Tsuna. (Courtesy of Kosumikku Shuppan)	160
Fig. 6.3	Image by Nonami Tsuna. (Courtesy of Kosumikku Shuppan)	161
Fig. 6.4	Image by Nonami Tsuna. (Courtesy of Kosumikku Shuppan)	163
Fig. 6.5	Image by Nonami Tsuna. (Courtesy of Kosumikku Shuppan)	164
Fig. 6.6	Image by Nonami Tsuna. (Courtesy of Kosumikku Shuppan)	166
Fig. 6.7	Image by Nonami Tsuna. (Courtesy of Kosumikku Shuppan)	167
Fig. 7.1	Image by Tanaka Keiichi. (Courtesy of KADOKAWA)	182
Fig. 7.2	Image by Tanaka Keiichi. (Courtesy of KADOKAWA)	194
Fig. 7.3	Image by Tanaka Keiichi. (Courtesy of KADOKAWA)	195
Fig. 7.4	Image by Tanaka Keiichi. (Courtesy of KADOKAWA)	197
Fig. 7.5	Image by Tanaka Keiichi. (Courtesy of KADOKAWA)	198
Fig. 7.6	Image by Tanaka Keiichi. (Courtesy of KADOKAWA)	200
Fig. 7.7	Image by Tanaka Keiichi. (Courtesy of KADOKAWA)	201
Fig. 7.8	Image by Tanaka Keiichi. (Courtesy of KADOKAWA)	204
Fig. 8.1	Image by Hosokawa Tenten. (Courtesy of Gentosha)	218
Fig. 8.2	Image by Hosokawa Tenten. (Courtesy of Gentosha)	229

Fig. 8.3 Image by Hosokawa Tenten. (Courtesy of Gentosha) 230
Fig. 8.4 Image by Hosokawa Tenten. (Courtesy of Gentosha) 232
Fig. 8.5 Image by Hosokawa Tenten. (Courtesy of Gentosha) 233
Fig. 8.6 Image by Hosokawa Tenten. (Courtesy of Gentosha) 235
Fig. 8.7 Image by Hosokawa Tenten. (Courtesy of Gentosha) 236
Fig. 8.8 Image by Hosokawa Tenten. (Courtesy of Gentosha) 238
Fig. 8.9 Image by Hosokawa Tenten. (Courtesy of Gentosha) 240
Fig. 9.1 Image by Shiramizu Sadako. (Courtesy of Seiwa Shoten) 250
Fig. 9.2 Image by Shiramizu Sadako. (Courtesy of Seiwa Shoten) 264
Fig. 9.3 Image by Shiramizu Sadako. (Courtesy of Seiwa Shoten) 265
Fig. 9.4 Image by Shiramizu Sadako. (Courtesy of Seiwa Shoten) 266
Fig. 9.5 Image by Shiramizu Sadako. (Courtesy of Seiwa Shoten) 268
Fig. 9.6 Image by Shiramizu Sadako. (Courtesy of Seiwa Shoten) 269
Fig. 9.7 Image by Shiramizu Sadako. (Courtesy of Seiwa Shoten) 271

CHAPTER 1

Introduction

Not a single day passes nowadays without the phrase *mental health* in the news. In this unpredictable time of the COVID-19 pandemic, our psychological well-being is negatively impacted by multiple factors, including social isolation, economic downturns, and a death, illness, or job loss in our families. It is of primary importance for us to confront the issue of mental health now. According to the World Health Organization's 2022 data, more than 100 million people are experiencing serious mental health problems in the Western Pacific Region, which includes the Philippines, China, and Japan.[1]

Before the pandemic, the government of Japan had already seen mental health as one of the nation's most urgent issues. According to the 2021 disability whitepaper issued by the Cabinet Office (*Naimushō*), 7.6% of the whole population of Japan (126.3 million people) have some form of disability.[2] By groups, 3.45% are people with physical disabilities; 0.87% are people with intellectual disabilities; and 3.32% are those with mental disabilities. The Cabinet's 2019 data shows that the increase in the number of people with mental disabilities is close to 1% in two years while the increased number of the other two groups is only about 0.1% in each.[3] The almost 1% increase in mental health cases marks a significant change in this pre-pandemic data.

Perhaps, to reflect this national urgency, a tsunami wave of change has taken over the field of graphic novels in Japan. Essay manga, the originally

Y. Okuyama, *Tōjisha Manga*,
https://doi.org/10.1007/978-3-031-00840-5_1

female-dominant, autobiographical comic genre of the 1980s in which popular artists wrote about their own lives as the subject, had proliferated between the 1990s and the 2000s. A subgenre of essay manga has emerged that has attracted significant media attention in the past twenty years: both male and female cartoonists narrate their direct experience with neurological or mental health problems in autobiographical comics. This book sheds light on the emergence and social impact of graphic memoirs about brain and mental health by analyzing five representative comic titles using semiotics and author interviews. With the current, worldwide, heightened importance of mental health, I feel that this particular consort of manga should be considered more relevant in our lives than ever.

In the absence of a label, I use the term *tōjisha manga* to refer to this increasingly noticeable offspring of autobiographical comics in Japan. Why use this special label? First, the label is necessary to distinguish this group from other autobiographical narratives that do not deal with such stigmatized conditions as psychological disability. In Japan's comics publishing industry, we have an established genre, *essay manga*, in which autobiographical works are classified. However, using that genre name feels insufficient because not all *essay manga* titles are self-portrayals of mental distress experience. No other word seems to capture this subgenre more aptly. Second, it appears felicitous because *tōjisha* is a semantically loaded word that is highly evocative of Japan's *tōjisha undō*, the human rights movements by minority groups. The word *tōjisha* also signifies the importance of *tōjisha kenkyū* (self-directed research conducted by people with disabilities), a topic that has been passionately debated in the field of disability studies in Japan. The key feature of *tōjisha manga* is that the authors themselves have dealt with the very condition illustrated in the stories.[4] I argue that it is a genre particularly suited to narrating one's mental health crisis by providing theoretical backgrounds for *tōjisha manga* such as the definition of the word *tōjisha* and a brief history of *tōjisha undō* movements in Japan as well as by examining various forms of first-person narratives, including life stories in the graphic platform, in the first half of this book. To demonstrate the humanizing power of *tōjisha manga*, I then present five case studies using representative manga titles in the latter half.

About This Book

The idea of writing about Japan's autobiographical comics began as an offshoot of the research designed for my previous book, *Reframing Disability in Manga*, funded by the Japan Foundation in 2017. The final chapter of that book included a graphic memoir by a transgender manga artist. I began to pay more attention to autobiographical comics then. At that time, I was unaware that the genre of *essay manga* existed in Japan. Wishing to explore graphic memoirs more, I began to conceptualize a new research project that focused on first-person narratives on mental disability, a disability category I did not examine closely in that book. However, the real seed for this monograph was planted in summer 2018 during my participation in *Global Histories of Disability*, a four-week summer institute of the National Endowment for the Humanities (NEH). There, I became interested in how mental disabilities had been documented throughout history in literature and other forms of narrative. There, I was also introduced to *mad studies*, a relatively new field of research and activism to fight the stigma of mental health problems in society. For years, I have been teaching a course on manga studies as well. Thus, to me, uniting both interests of scholarship seemed to be a perfect marriage. During the Fourth of July break from the NEH program, I sent out book proposals and eventually chose Palgrave Macmillan as my publisher.

In the same year, I also had an opportunity to take a twelve-session course organized by the National Alliance on Mental Illness (NAMI). For decades, I have been the primary caretaker of my mother who was diagnosed with obsessive-compulsive disorder (OCD). Her condition had taken an emotional and financial toll on our family life since I was very small. It was not easy to recover from both the psychological and financial setbacks caused by this illness over the years. In that sense, I am a survivor of caregiving. Because of my lifelong experience with her OCD, I intentionally avoided the topic of mental disability while writing my previous book on disability. However, once I made an intellectual decision to confront that personal history, everything fell into place with the impetus to grow the seed into a book in the following two years. After the NAMI classes, the NEH program, and the book contract came my participation in a panel discussion at the spring 2019 Association for Asian Studies (AAS) annual conference in Denver, Colorado. Through the panel, I was fortunate to work with two prominent scholars specializing in Japanese disability studies, Karen Nakamura, author of *A Disability of the Soul: An Ethnography of Schizophrenia and Mental Illness in Contemporary Japan* (2013), and Caroline S. Stevens, author of *Disability in Japan* (2013). That panel and other venues allowed me to submit this research project for scholarly feedback, which subsequently shaped my interpretation of *tōjisha*, my justifications for the coinage of *tōjisha manga*, and the like. As stated in the acknowledgments, I received a seed-money grant from my institution and another research grant to conduct fieldwork, including interviews with several manga artists, in Japan in 2019 and 2020. Unfortunately, the pandemic interfered with my planned fieldwork and time secured for writing. However, thanks to the support and flexibility of the individuals and organizations, especially my editor and her assistant at Palgrave Macmillan, I was able to resume this project by summer 2021. For the final manuscript, I selected five artists and their amazing works from all the individuals I interviewed. Therefore, besides some setbacks, milestones and breakthroughs have occurred, one after another like a force of gravity, for the creation of this book.

Purpose and Target Audience

The main purpose of this book is to examine selected works of *tōjisha manga*, Japan's autobiographical comics of literary gravitas in which the author's experience of a mental health problem is recounted in a unique medium that combines text and image for storytelling. I use Japanese comics as a form of popular media to exemplify the significance of representing authors' lived experience in graphic memoirs, an act that was traditionally eschewed in the world of comic publications. My particular focus of text analysis is on each artist's unique strategies to express graphically the emotional highs and lows of such a lived experience, while showing the reader that a mental health problem is something rather relatable and ubiquitous in society.

This book is different from my previous book on disability seen through the lens of comics in three measures. First, it focuses specifically on autobiographic comics, not the traditional, mainstream genre of story manga. Second, unlike the more noticeable disabilities such as paraplegia and blindness explored in my previous book, this book analyzes manga titles that narrate the experiences of "invisible" mental disabilities since health issues brought by neurological or psychological conditions differ from most cases of physical disabilities in that symptoms such as a heightened sense of anxiety or having thoughts about suicide are not easily noticeable to others. This invisibility contributes to the fact that diagnoses such as a panic disorder or depression are left out of the public eye, a situation that can be a double-edged sword. While it is easier to hide the condition and thus temporarily possible to avoid stigma, it may bring prejudice and misunderstanding in effect. For instance, one's personality, ability, or attitude tends to be blamed for seemingly antisocial behaviors prompted by a mental distress. Third, the most distinctive difference is this book's detailed semiotic analysis of comics' contents, including manga-specific signifiers embedded in selected panels.

The primary target audience for my book are scholars of Asian popular culture, especially those who are interested in Japanese comics, particularly the genre of autobiographical comics of stigmatized health conditions such as depression. Humanities scholars in Western comics studies and media studies may also benefit from reading this book as no such book has been published for English-language readers to the best of my knowledge. The book is also intended to augment scholarly analyses of graphic novels such as *The Visual Language of Comics* by Neil Cohn, *Graphic Medicine Manifesto* by MK Czerwiec et al. (2015), *Alternative Comics* by Charles Hatfield (2006), *Comics and Narration* by Thierry Groensteen and Ann Miller (2013), and *An Introduction to Manga* by Fusanosuke Natsume and Takeuchi Osamu (2009). I hope that it also contributes to the field of comics studies in the genre of autobiographical comics on serious social issues such as disability, gender, and medicine cultivated by North American and European academics, including Andrew J. Kunka and Elizabeth El Refaie. That said, this book is also open to a broad scholarly readership, including undergraduate students who are interested in manga and its societal contributions in Japan. Therefore, I chose a scholarly commercial publisher and kept the book accessible to general readers without diminishing scholarly rigor.

Basics of Manga and Methodology for Manga Analysis

It is now fairly well understood that *manga* is a Japanese word for comics and is also translated as Japanese graphic novels. However, the basics of manga may not necessarily be known to many English-language readers. Therefore, in this section, I provide a brief description of this medium.

English-language papers may use the phrase *graphic novels* in reference to comics that are of literary gravitas and deal with the social and ideological issues faced by people with mental disability. In Japan, however, comics is always called manga regardless of the characteristics of a story. Manga is a form of art with its own visual structure, not only different from other art forms such as photography or painting but also from other types of comics in the world. Manga is typically created using black-and-white, still drawings (except for feature pages) along with accompanying dialogue. However, recently, more and more digitalized manga have become full-color manga (e.g., Takagi Naoko's 2013 essay manga, *Marason ichinensei*, and Sato Miyuki's 2017 essay manga, *Haha to ironna tokoro ni ittekimashita*). Unlike anime, manga tell their stories through still images only, no moving pictures. While US mainstream comics are represented by two major comics companies, DC Comics and Marvel Comics (Chute 2017), numerous companies publish manga of various genres in Japan (e.g., Kōdansha, KADOKAWA, Seiwa Shoten, Komikku Shuppan, Gentōsha, and Shōgakkan, to name a few).

Another concept helpful to understanding manga is *mangaka* (manga artists), who both draw and write their stories. The Japanese suffix *-ka* refers to a title given to someone specialized in a profession such as *geijutsuka* (artist). On occasion, the creation of a manga series may involve another individual, *gensakusha*, the original writer of a story or a writer hired to create the storyline for a *mangaka* (e.g., *Boku no kanojo wa hattatsu shōgai*, or *My Girl-friend Is an Aspie*, a two-volume essay manga written by Kurage and illustrated by Terajima Hiro 2013). Although many works that first debut on digital comics sites and subsequently gain popularity go directly into comic book format nowadays, most manga stories are initially serialized in a magazine devoted to a certain readership.

Furthermore, manga comes in different genres, including *shōjo manga*, *shōnen manga*, *gakushū manga*, *gag manga*, and *essay manga*. For example, *shōjo manga* refers to a genre for young female readers, while *josē manga*, also known as *rediisu komikku* (ladies' comics), is a genre developed during the 1980s that targets a more mature female readership. The genre of *shōnen manga* was developed for teenage boys, while *sēnen manga* was created for young adult and adult males. However, distinctions between genres have recently become blurred (e.g., *Shingeki no Kyojin*, or *Attack of the Titans*, a manga that is popularly read by both adolescent and adult male readers and can be either *shōnen* or *sēnen manga*). Furthermore, *story manga* is not a new genre but is a term used by Japanese manga scholars such as Natsume and Takeuchi (2009) to collectively refer to traditional, mainstream manga (e.g., *shōjo manga* and *shōnen manga*) to distinguish from other, more specialized types such as *gag manga* and *essay manga*.

Manga basics also include its reading directionality as the Japanese sequence of comic frames is arranged differently from that of English-language comics. The general rule of manga is reading from right to left and from top to bottom, which is based on the traditional vertical reading style in Japan and China. Thus, in Fig. 1.1, an image taken from Hosokawa

39 PART 1 ある日、突然、うつ病に。

Fig. 1.1 Image by Hosokawa Tenten. (Courtesy of Gentosha)

Tenten's manga, *My SO Has Got Depression*, the first frame to read is the top-right frame (follow the arrows for the correct reading directionality).[5] Another important aspect of reading manga is that words are written vertically in speech or thought balloons as shown in Fig. 1.1. The Japanese reader's eyes naturally move from top to bottom, right to left. In this example, the reader will look at the right-most balloon in the top-right panel first and then move to the middle balloon in the same panel.

Both visual and textual elements are imbued in manga. The textual information consists of words written in balloons and around the images in the panel space or the margin as well as words indicating sound and manner effects such as the handwritten word "Hat!" (a signifier that indicates the character's sudden realization) in the second panel and "*Buchi butsu buchi butus buchi butsu*" as a symbolic sound of complaint. The semiotic system of manga includes *onyu* which are written words that function as sound effects (similar to onomatopoeia) or represent manners and emotions as well as *manpu* (also called *keiyu*) that are drawn symbols implying the scene's atmosphere or the character's inner state. In this book, manga analysis includes the examination of how such signifiers are used for what effects. Therefore, manga reading can be complicated by the ways in which multiple elements, including panels, speech balloons, other text, and *onyu* and *manpu* signifiers, interact with one another to tell a story. Some of these elements are semiotic signs common in comics in general (e.g., Ranker 2014), while others are manga-specific signifiers that govern this communication system akin to grammar in language (e.g., Cohn 2010, 2013).

Manga analysis in this book has two parts: the micro-level analysis via manga semiotics and macro-level of sociocultural factors influencing each comic narrative. The micro level focuses on semiotic compositions embedded in each graphic memoir. I chose semiotics as an analysis approach primarily because of semiotics' strengths in illuminating graphic metaphors of pain, panic, and other emotional outbursts and demonstrating how the comic artists made these invisible signs visible. I deem semiotics to be the most appropriate tool to examine the visual text of manga. Therefore, the analysis chapters showcase how the artists interspersed the heavy topics of stigma, self-shaming, and sometimes suicidal ideation with lightweight slapstick moments in their autobiographical comics. As for the macro-level analysis, each analysis chapter provides the background of the author and the genesis of their manga story obtained from my interviews with the artists. Furthermore, their comments are also analyzed with sociocultural factors about the particular condition in discussion (e.g., how depression is misconceived by the lay Japanese). All the interviews are edited for length and clarity and interspersed with scholarly information to explain brain or mental health issues. Audience response analysis and other methods are also vital in understanding the social influence of autobiographic comics. Therefore, data about the selected manga such as awards and recognitions are included.

A Word About Language

Various terms of self-identification exist for people with mental health issues in Western contexts, including psychiatric system survivors. Recently, the term *mad* has been used in reference to psychiatric conditions. The term *madness* began to be applied as a way of self-identifying with the mental patient liberation movements in the 1960s and is now used as an umbrella term to represent a diversity of these identities associated with mental illness, including other labels such as mental health consumers and ex-patients (Menzies et al. 2013: 9). As opposed to scientifically approved terms such as *mental illness*, the word *madness* carries a political implication, facilitating "a connection between academic critiques and the Mad Pride movement of activists, artists, and intellectuals who have followed in the footsteps of Black Pride and Gay Pride to destabilize and reverse the binaries and hierarchies associated with mainstream psychiatry" (Lewis 2009: 155). The term is also evocative of Michel Foucault's *Madness and Civilization: History of Insanity in the Age of Reason* (1961) and has a certain open-endedness to it that allows authors of psychiatric memoirs to more progressively discuss their alternate states of consciousness without the linguistical restriction of scientific terms. Used in academic contexts mostly in Australia, England, and North America, the word *mad* is a label that connotes "positive and person-centered discourses" (Price 2011: 9).

Besides the word *madness*, several contemporary labels of mental health exist in the English language, including *psychiatric disability*, *mental illness*, and *neurodiversity* as conditions, and *the neuroatypical*, *mental health service users*, and *psychiatric system survivors* for people. Although scholars of disability studies carefully select a term for their books out of social considerations (e.g., *mental disability* in Price's *Mad at School*, 2011), how to name mental disability is always a matter of contention. Therefore, I owe the reader an explanation for what language I use and why I use it in this book. First, I attempt to avoid pathological terms like mental illness and disorders as much as possible. Even though they may be properly used in medical and government documents about mental disability, those words are not value free and objective in other contexts and may further reinforce the stigma surrounding mental health conditions. In consideration of this, I elected to use more neutral expressions such as mental health problems, neurological differences, emotional distress, mental-ill health, and persons with lived experience. On occasion, I use mental disability as a categorical term. Second, as discussed in chapter "Tōjisha", there is an all-inclusive term for people with mental disability: *tōjisha*. Once the background of the term is sufficiently presented, the word *tōjisha* appears as frequently as the non-medicalized expressions listed earlier. Third, primarily in the case study chapters, I move beyond those terms to use specific labels applied in the World Health Organization's *International Classification of Diseases*, 11th Edition (ICD-11 2022), as well as the American Psychiatric Association's *Diagnostic and Statistical Manual of Mental Disorders*, 5th Edition (*DSM-5*). That is partly because the interviewed manga artists used the medical labels themselves and also because I have to stick with a label consistently as I describe the condition's symptoms and diagnoses.

I believe that we should see the individual before their physical or mental condition. However, due to literary and editorial considerations as well as the *tōjisha*'s own wording (e.g., the word *byōki*, or illness, used by some of the interviewees), more medical terms had to be applied. All in all, it is my hope that the correct tone has been set here and that I speak sensitively for the *tōjisha* in my use of language throughout this book.

Selection of Manga Titles and Mental and Neurological Conditions

Titles chosen for this book are autobiographical comics of literary gravitas. As I pointed out in my previous book on manga, the media's overall failure to accurately present disability has been criticized by academics, especially scholars in the fields of media studies and disability studies. Although disability is more seriously and carefully represented in mass media nowadays, the "unreliable narrator" stereotype is still reinforced in the portrayal of the protagonist with psychiatric symptoms through contemporary literature, film, art, theater, and comics where discriminatory words such as *crazy* are sometimes applied in colorful ways. For this reason, I carefully selected comic titles as representative *tōjisha manga* for this book. Thanks to the growing number of autobiographical comics on the topic of mental health, I was able to find approximately ninety *tōjisha manga* titles and interview several authors from which I selected a handful of masterpiece stories for the limited space of this book: *Gaki no tameiki* (Okita Bakka), *Akira-san wa asuperugā* (Nonami Tsunami), *Utsu nuke* (Tanaka Keiichi), *Tsure ga utsu ni narimashite* (Hosokawa Tenten), and *Uchi no OCD* (Shiramizu Sadako). I included memoirs authored by spouses because they are family/caregiver *tōjisha* and their voices augment the "insider" perspectives of the lived experience of mental disorders. More detailed justification for including the graphic memoirs by caregivers is provided in chapter, "Tōjisha," in Part I. The chapters that examine the works by direct *tōjisha* authors (Okita, Nonami, and Tanaka) precede those by caregiver *tōjisha* authors (Hosokawa and Shiramizu) in Part II.

Each of the selected manga titles features either a psychological (e.g., depression, schizophrenia) or a neurological condition (e.g., Asperger's syndrome; Attention Deficit Hyperactivity Disorder, or ADHD). Although serious illnesses represented in recent autobiographical comics include cancer, Parkinson's disease, and hyperthyroidism and are equally worthy of attention, I decided to focus on mental disabilities that have been affecting millions of young adult and adult populations in Japan and beyond. Along with the aforementioned bases for the selection of mental disorder categories, I chose to keep a narrow focus on these types of mental disability, especially because of recent media coverage of mental health issues not only in Japan but also in other developed countries. Specific medical categories featured in my case studies of *tōjisha manga* are *hattatsu shōgai* (developmental disorders, including ADHD), the hottest topic in both academic research and popular culture in Japan; Cassandra syndrome, a newly debated mental health condition known to afflict neurotypical partners of Aspies; *utsubyō* (major depression), which is the most diagnosed psychological disorder in Japan; and *kyōhakusei shōgai* (OCD), one of the least understood and personally the most relevant topic. Along with the aforementioned bases for the selection of these four conditions, I chose to keep a narrow focus on them especially because of recent media coverage of mental health issues in Japan. However, related mental health issues such as panic attacks and Asperger's syndrome are also described if they are experienced by the interviewed artists.

Following the pattern of approachable academic books such as Geoff Rolls's (2015) *Classic Case Studies in Psychology*, I opted to cover this small set of conditions more intensely rather than a wide range of psychological disabilities. Of course, schizophrenia and other psychotic illnesses are also serious and stigmatized medical conditions. Dissociative identity disorder (DID), as seen in the film *Split* (2016), is one of the most demonized mental disabilities in popular culture, an unfortunate state of affairs that needs to be corrected. Furthermore, eating disorders, self-harming, and suicidal ideation are so prevalent among adolescents and young adults that I feel manga about these deserve academic attention. And yet all these could not be covered within the book's limited scope. In describing these conditions, I referred to the classifications in *DSM-5* and ICD-11. Neither I as the author nor the reader is expected to know the whole list of medical conditions in those classifying systems.

As mentioned earlier, this book draws from both ICD-11 and *DSM-5* as the main sources of information. ICD-11 was released in 2019; by comparison, *DSM-5* became available in 2013. *DSM-5* is used more predominantly in the United States where I live and is a closer match to the types of disorders featured in this book. In comparison, ICD-11 covers a much more diverse list of illnesses. However, both coding systems are influential diagnostic sources. Furthermore, Japanese psychologists and psychiatrists draw from both classification systems in discussing diagnoses. Therefore, I decided to refer to both, especially for an audience beyond the United States, and draw diagnostic information from either *DSM-5* or ICD-11.[6]

Since I began to pay attention to *tōjisha manga* in the summer of 2018, the volume of this subgenre has been growing. As mentioned earlier, approximately ninety units of such manga—some are single titles while others are a series—have been published in Japan as of 2022. Due to the scope of this book, only a fraction of them are introduced and analyzed here. To augment, I will continue to publish through academic journals and other venues my ongoing research on *tōjisha manga* that feature other serious and socially stigmatized conditions such as dissociative identity disorder, eating disorders, and anxiety disorders (e.g., Tokin's *Jitsuroku Kairisei-shōgai no chiguhagu na hibi* for dissociative identity disorder and Nagata Kabi's *Sabishisugite rezu-fūzoku ni ikimashita repo* for eating disorders and self-mutilation).

The Structure of the Book

The book is structured in two parts: Part I "Foundations" and Part II "Case Studies." In Part I, the first chapter lays the book's foundations by introducing the key word *tōjisha* as the background of the subject *tōjisha manga*. Through the literature review, chapter, "Tōjisha," explains what the word originally meant, who are considered *tōjisha* in this book, how the *tōjisha* movement influenced Japan's attitudes toward people with disabilities, and how the groundbreaking perspectives of *tōjisha kenkyū* came into being for people with mental disability, in particular. The chapter answers questions such as who a *tōjisha* is and why lived experience matter.

The second chapter discusses topics associated with first-person narratives, from the nature and aims of autobiography to the recent emergence of life writing on disability and illness. The chapter's main focus is on the empowering effects of *tōjisha* narratives, memoirs by psychiatric system survivors in Japan. It ends with a brief introduction of graphic memoirs as a way to segue into the next chapter, which covers the historical and sociocultural background of *essay manga*, autobiographical comics in Japan. Because the majority of *tōjisha manga* are created in the style of this genre, I decided to have a single chapter of *essay manga* as chapter, which features the topic of *essay manga*, a genre of autobiographical comics that emerged in the 1980s and are becoming increasingly popular in the world of social media in Japan. These comics tell the author's daily occurrences and real-life experiences. This chapter describes how *essay manga* convey the author's perspectives on serious themes such as mental health issues and child-rearing difficulties via first-person narration and teach us the wisdom of life through the healing power of humor and entertainment. Both chapters "Tōjisha Narratives" and "Essay Manga" provide a thorough discussion of relevant topics that helps contextualize the case studies to be introduced in Part II. Both chapters address key questions while defining important terms and concepts that appear recurrently throughout the book. Questions addressed and responded to in these chapters include: Why should we pay attention to first-person narratives now? What factors have contributed to the rise of *essay manga* in Japan? What makes autobiographical comics special?

In Part II, all five chapters feature the analysis of selected *tōjisha manga* as case studies, examining how psychological or brain-related symptoms are artistically depicted in representing panels. Each case study is presented to argue that *tōjisha manga* narrative is more authentic than that of story manga and empowering precisely because it comes directly from the lived experience of the *tōjisha*, not from medical professionals or disability studies scholars. Each case-study chapter semiotically analyzes the artist's visual and other techniques while providing the manga's genesis, the artist's background, and information about the book's reception. Questions addressed include: What visual strategies are used by the authors of the introduced *tōjisha manga* to represent invisible matters such as symptoms? What did these authors accomplish in order to destigmatize the condition as they narrate a highly personal, life-altering experience?

As the first chapter of Part II, chapter "Okita Bakka's *Gaki-Tame* Series (2011–2013): A Memoir of a "Troublemaker" Aspie Girl" introduces Okita Bakka's *tōjisha* manga titled *Gaki no tameiki*, which features Okita's lived experience with *hattatsu shōgai*. This is a visually innovative graphic autobiography about the manga artist's days as an elementary schoolgirl who was constantly labeled a "problem child" by teachers. Currently, *hattatsu shōgai* is one of the most media-popular and academically researched mental health topics, besides depression, in Japan. That is one of the reasons why I included this topic in this book. Chapter "Nonami Tsuna's *Akira-san* Series (2011–2017): A Memoir of a "Cassandra" Wife" showcases Nonami Tsuna's eight-volume series, *Husband, Akira, Has Asperger's Syndrome*. After briefly describing husband Akira's adult Asperger's syndrome, the chapter presents, in detail, wife Tsuna's Cassandra syndrome, a mental health phenomenon experienced by people in a very close relationship with someone with Asperger's syndrome. This syndrome has been featured in many women's magazines, and more *tōjisha manga* on this condition have been published in Japan. Chapter "Tanaka Keiichi's *Utsunuke* (2017): An Ode to Depression Tōjisha" analyzes Tanaka Keiichi's multicolor graphic memoir about his lived experience of depression. Having struggled with depression for approximately ten years, Tanaka imbues both visually and textually personalized messages for his fellow tōjisha in his manga *Utsunuke: Utsu tonneru wo nuketa hitotachi*. In contrast, chapter "Hosokawa Tenten's *Tsure utsu* Series (2006–2013): A Couple's Lived Experience of Depression" presents a tōjisha narrative about depression from the perspective of a tōjisha's wife, using Hosokawa Tenten's essay manga, *Tsure ga utsu ni narimashite*. Chapter "Afterword" concludes Part II with Shiramizu Sadako's *Uchi no OCD*, a graphic memoir narrated similarly in a caregiver perspective but about a different mental health topic. Shiramizu's story tells the reader about an uncanny aspect of family life affected by OCD, a serious mental health issue manifested by obsessions and compulsions, which thousands of people in Japan still hide in public to this day.

Tōjisha manga is a unique subgenre of essay manga that draws on real-life narratives created by the *tōjisha* themselves. Their stories are serious-yet-humorous tales of how they have coped with serious conditions such as depression and schizophrenia. A scholarly report of this emergence in the comics world needs to be available to the English-language readership. Furthermore, Japan's issues of *tōjisha* can relate to the disability rights motto that resonates throughout the world: "Nothing about us without us." Understanding the individuals' points of view and their unique experiences of a stigmatized mental disability is critically important as they are quite often misunderstood by the public as having "personality flaws," being totally unreliable, or of using their condition as an "excuse" to get what they want. I argue that *tōjisha manga* is a significant and relevant genre that is worthy of scholarly examination because it narrates the distress, isolation, and low self-esteem that stem from societal reactions to such "invisible" disability. Exemplified by *tōjisha manga*, this new movement of self-empowerment should be introduced to Western scholars of comic studies and disability studies as Japanese popular culture can now offer the new perspective of mental disability discourse. Most importantly, this book captures many helpful insights shared by the interviewed artists of indomitable spirit, which in and of itself is valuable to many readers especially during the pandemic that impacts our mental well-being.

Notes

1. https://www.who.int/westernpacific/health-topics/mental-health.
2. In Japan, the government data on disability are presented in three categories: "people with physical disabilities," "people with intellectual disabilities," and "people with mental disabilities." The Cabinet Office has been publishing a disability whitepaper (*shōgaisha hakusho*) annually since 1994.
3. https://www8.cao.go.jp/shougai/whitepaper/r03hakusho/zenbun/siryo_02.html.
4. It is important to clarify that although this book limits itself to examining graphic memoirs in the category of mental disability, manga titles by *tōjisha* with physical disabilities do exist albeit their number is quite small. Examples are works by Matsutani Takuya, a deaf manga artist; Banii Urasawa, a visually impaired manga artist; and Sashida Hiroyuki, a manga artist who was first hospitalized for a cerebral tumor and then became incapacitated from a hit and run four years later. All three have published their comic works based

on their tōjisha experiences. However, this group of manga is far more limited compared to the growing number of manga on mental disability.

5. Although manga appears to be read in Japanese format, that characterization is not entirely accurate. More traditional writings such as novels and newspapers are read from top to bottom and right to left; however, scientific and educational documents, including academic journal articles and textbooks, are typically read from left to right just like English sentences.
6. ICD-11, which officially came into effect in January 2022 (https://icd.who.int/en/), was not yet cited in academic papers and other scientific references during most of my manuscript writing. Thus, I drew information more from *DSM-5*, especially in the chapters about depression. Furthermore, it was only recently that the United States began evaluating the new edition for adaptation.

References

Chute, Hillary. 2017. *Why Comics?: From Underground to Everywhere*. New York: HarperCollins.

Cohn, Neil. 2010. "Japanese Visual Language: The Structure of Manga." In *Manga: An Anthology of Global and Cultural Perspectives*, edited by Toni Johnson-Woods, 187–203. New York: Continuum Books.

———. 2013. *The Visual Language of Comics: Introduction to the Structure and Cognition of Sequential Images*. London: Bloomsbury Academic.

Czerwiec, MK; Michael J Green, Susan Merrill Squier, Kimberly R Myers, Scott T Smith, Susan Merrill Squier, and Ian Williams. 2015. *Graphic Medicine Manifesto*. University Park, PA: Penn State University Press.

DSM-5 (*Diagnostic and Statistical Manual of Mental Disorders, 5th edition*). 2013. Washington, DC: The American Psychiatric Association.

Foucault, Michel. 1961. *Madness and Civilization: History of Insanity in the Age of Reason*.

Groensteen, Thierry and Miller, Ann. 2013. *Comics and Narration*. Jackson, MS: University Press of Mississippi.

Hatfield, Charles. 2006. *Alternative Comics: An Emerging Literature*. Jackson, MS: University Press of Mississippi.

ICD-11 (*International Classification of Diseases, 11th Edition*). 2022. Geneva: The World Health Organization.

Kurage and Hiro Terashima. 2013. *My Girlfriend has Developmental Disorder* (*Boku no kanojo wa hattatsu shōgai*) (1). Tokyo: Gakken Purasu.

Lewis, Bradley. 2009. "Obsession: A History; The Loss of Sadness: How Psychiatry Transformed Normal Sorrow into Depressive Disorder; Shyness: How Normal Behavior Became a Sickness, and Bipolar Expeditions: Mania and Depression in American Culture (review)." *Literature and Medicine*, 28 (1): 152–171.

Menzies, Robert; LeFrançois, Brenda A.; and Reaume, Geoffrey. 2013. "Introducing Mad Studies." In *Mad Matters*, edited by Brenda A. LeFrancois, Robert Menzies, and Geoffrey Reaume, pp. 1–22. Toronto: Canadian Scholars' Press.

Nakamura, Karen. 2013. *A Disability of the Soul: An Ethnography of Schizophrenia and Mental Illness in Contemporary Japan*. Ithaca, NY: Cornel University Press.

Natsume, Fusanosuke and Takeuchi, Osamu. 2009. *An Introduction to Manga*. Tokyo: Minerubashobo.

Price, Margaret. 2011. *Mad as School: Rhetorics of Mental Disability and Academic Life*. Ann Arber, MI: The University of Michigan Press.

Ranker, Jason. 2014. "The emergence of semiotic resource complexes in the composing process of young students in a literacy classroom context." *Linguistics and Education*, 25: 129–144.

Rolls, Geoff. 2015. *Classic Case Studies in Psychology*. New York: Routledge.

Sato, Miyuki. 2017. *Haha to ironna tokoro ni ittekimashita*. Tokyo: Saizusha.

Stevens, Caroline S. 2013. *Disability in Japan*. New York: Routledge.

Takagi, Naoko. 2013. *Marason ichinensei*. Tokyo: KADOKAWA.

PART I

Foundations

CHAPTER 2

Tōjisha

In Western contexts, various terms of self-identification exist for people with mental health issues, including *psychiatric system survivor*. Other labels such as *mental health consumer* and *ex-patient* are also in use (Menzies et al. 2013: 9). Margaret Price, author of the seminal work, *Mad at School* (Price 2011), further explains that the word *mad* is a label "generally used in non-US contexts" and connotes "positive and person-centered discourses" although it is "less recognizable in the United States" (9). Listing contemporary mental health labels used in the United States such as *psychiatric disability*, *mental illness*, *cognitive disability*, *mental health service user (or consumer)*, *neurodiversity*, *neuroatypical*, *psychiatric system survivor*, *crazy*, and *mad*, she elects to use the term *mental disability* in that book but warns the reader that "the problem of naming has always preoccupied" scholars of disability studies (9).

In Japan, people with a disability are officially referred to as *shōgaisha*, and the government classifies people with disabilities into three groups of *shōgaisha*: *shintai shōgaisha* (the physically disabled), *chiteki shōgaisha* (the intellectually disabled), and *seishin shōgaisha* (the mentally disabled). Because of the negative connotation of the *kanji* character for *shōg*ai (obstacle, hindrance), suggestions for alternative labels have been proposed (e.g., replacing the *kanji* with its phonetic version in hiragana). However, people who have a disability have elected the word *tōjisha* instead as the preferred self-referential term. This chapter explains what it

Y. Okuyama, *Tōjisha Manga*,
https://doi.org/10.1007/978-3-031-00840-5_2

means to be a *tōjisha* and why I have chosen this term to refer to the authors of the autobiographical comics analyzed in the latter part of this book. The chapter also responds to the following central questions germane to the coinage of *tōjisha manga*: Who constitutes the members of *tōjisha*? For example, can a primary caregiver be a *tōjisha* and legitimately speak of the lived experience of a person with a mental health problem? Why is it important to hear *tōjisha* voices? In addition, this chapter explores concepts associated with *tōjisha* such as *tōjisha mesen* (their own perspectives) through the literature review of disability rights movements in Japan and compares the philosophy behind *tōjisha* in Japan's *tōjisha kenkyū* (self-research) in the context of mental disability with the idea of service users propagated in mad studies in North America.

Because the term *tōjisha* appears quite frequently in this chapter, it will be unitalicized hereafter unless it is used as part of another, new technical expression such as *tōjisha undō*.

Who Is a Tōjisha?

Originally used in legal parlance, the Japanese word tōjisha consists of *tōji* (being directly concerned) and *sha* (parties) referring to insiders involved in litigation. The word has evolved to mean the membership of a minority population. In essence, a tōjisha refers to any minority, including people with a mental or neurological disability, whose voice has been suppressed in mainstream discourse (Nakanishi and Ueno 2003/2015; Ueno 2009; Ishihara 2013, 2015). Throughout its evolution, the term tōjisha has accrued multiple meanings in various sociocultural fields, one of which is disability studies. Even though a referent of the term is almost as ambiguous as the word *person* or *agent* in public discourse, it is important to underscore that users of mental health services refer to themselves as tōjisha. How has the term come to be used in that context? This section provides a theoretical background in which I have elected to apply tōjisha for the book title instead of the government's official term, *seishin shōgaisha* (the mentally disabled).

Japanese anthropologist Teruyama (2011) translates tōjisha rather literally as "those who are concerned." According to Teruyama, the term has been a key concept of the Japanese disability rights movement since the 1970s and has helped activists develop identity and agency in the movement. Although the term is criticized for its selective nature and for failing to unite the diverse population involved in disability issues, people with disabilities were able to fight for their needs and rights under the banner of tōjisha. By contrast, American anthropologist Nakamura (2013) translates the word tōjisha as "people with disabilities or families thereof" (274). Although her book provides a similar historical account for the semantic shift from the legal term to the social movements' buzzword, what distinguishes Nakamura's view is her added description that the term is "used as a polite way within discriminated communities to indicate membership" (173). For example, supporters can respectfully address members of a minority class as *tōjisha no kata* (Mr. or Miss. Tōjisha).

McLelland (2009) offers a detailed historical account of the term's semantic shifts as follows. Before World War II, the word tōjisha "originated in the fields of law, politics, and administrative studies; has been used continuously in these disciplines across the course of the twentieth century"; and means "people or parties directly involved in the matter, often the matter of a litigation/lawsuit" (195). The term was applied more broadly in the women's liberation movement first and then in the fields of social welfare and disability studies as an inclusive term for minority groups in Japan in the early 1970s. A wide range of minority groups pioneered *tōjisha undō* (movements by and for the tōjisha) in the early 1980s, and their coalitions were formed as *tōjisha gurūpu* (groups of tōjisha). McLelland (2009) states, however, that although such movements by different coalitions of minority groups were unprecedented, the tōjisha groups failed to achieve strong solidarity due to internal conflicts.

The use of the term tōjisha afforded more agency to people belonging to a coalition. In the context of law, the perspective of the tōjisha is more privileged because those who are directly involved in a court case are more impacted by the lawsuit than a third party. Thus, by characterizing others as *hi-tōjisha* (non-tōjisha; those who are not directly concerned), the tōjisha themselves gain an advantage to push forward their viewpoints. This empowerment of the tōjisha, however, has a drawback: McLelland (2009) argues that the *hi-tōjisha* gets pushed aside as those who "cannot claim to know the impact of discrimination" (196). While it is a positive outcome in that the needs of directly impacted individuals are heard and favored in tōjisha movements, unlike previous generations, this dichotomy of *tōjisha* vs. *hi-tōjisha* has become a point of contention in public discussions. Another stumbling block is that it has become difficult to determine who exactly the tōjisha are and whose point of view is more privileged. Therefore, arguments centering around the concept of *tōjisha sei* (tōjisha-ness, or what makes one a legitimate tōjisha) continue endlessly, debating who is allowed to speak for whom.

The issue of *tōjisha sei* was vehemently debated by scholars, including the well-known sociologist and feminist Ueno Chizuko, in the 1990s. Ueno (2009) defines tōjisha as parties who knowingly accept support services. She argues that the tōjisha must have awareness of who they are and what their needs are. In other words, they actively identify with a group to which they believe they belong and share the same needs. She defines the nature of tōjisha in even more detail; for example, if a battered wife is not self-aware that she is affected by her spouse's domestic violence (DV), she is not yet a "tōjisha." Only when she gains a clear idea of what her needs are as a DV victim does she become a tōjisha.

In *Tōjisha Sovereignty* (Nakanishi and Ueno 2003/2015), Ueno's seminal work coauthored with the disability studies scholar Nakanishi Shoji, the tōjisha are people whose needs are not met by society (9). Nakanishi and Ueno emphasize the importance of the tōjisha's subjectivity, pitting it against the impersonal objectivity of outside authorities such as medical experts. The authors proposed *tōjisha gaku* (tōjisha studies) to establish theories about minority experience and voices and to delineate the concept of tōjisha once again. They argue that the tōjisha are the end users of services—medical, legal, or whatever services they need. In their opinion, anyone is potentially a tōjisha once such a need arises. However, if the situation is adjustable and their need subsides, then they are not tōjisha. Only if they are strongly aware that their situation cannot be amended unless their needs are met by certain services, are they tōjisha (as an example, interpreting services for deaf people's communication needs). Drawing from the notion of human dignity, Nakanishi and Ueno proclaim the sovereignty of the tōjisha (*tōjisha shuken*) in the case of the women's lib movement (*josei kaihō undō*) in Japan, as an example. They assert that so-called women's issues are actually societal problems imposed on women; it is not women but society that caused such problems. Sovereignty means that people can, and should, decide what to do for themselves because they know what is best for themselves. Thus, *Tōjisha Sovereignty* rejects paternalism. When this paradigm shift occurs in the consciousness of oppressed individuals, they become tōjisha, ones who have needs (10–11).

From the *Tōjisha Sovereignty*'s point of view, such neglected people include, but are not limited to, women, children, the elderly, disabled individuals, sexual minorities, and people with mental disability, all of whom inherit problems that society imposed. More closely in connection with disability studies, Fujii (2014/2016) explains that the idea of *Tōjisha Sovereignty* for people with disabilities originated in the Independent Living (IL) movement (*shōgaisha jiritsu-seikatsu undō*) propagated by disability rights activists in Japan, especially the members of the Association of Green Grass (*Aoi shiba no kai*) in the 1970s. The IL movement prompted the government to establish a support service system (*shien-hi seido*) in 2003for people with disabilities. The system now enables people with disabilities to choose the type of services they need without a third party's intervention. In Japan, the rate of living independently is still quite low for disabled individuals: people in their twenties (90% of them living with their families), thirties (80%), forties (60%), fifties (30%), and sixties and older (16.5% with their parents in their nineties) according to Fujii. Still, that movement did inspire many disability communities to take matters into their own hands because, at that time, the tōjisha were deprived of the right to decide what to do about their own issues.

Fujii (2014/2016) points out that not everyone has equal access to disability services; who gets what service largely depends on the type of disability. He gives an example in which people with Asperger's syndrome and neurodegenerative disorders are not included in many disability-related policies. Under these contexts, the word *tōjisha* is strongly associated with disability studies and the disability rights movement. Regarding the term's connection with social welfare and disability, McLelland (2009) also explains that when the United Nations designated 1981 as the International Year of Disabled Persons, people promoting the disability self-support movement adopted the label and called themselves tōjisha. To gain support and approval from society, the activists of the movement asserted their needs, including the right to make decisions on their own, because they were none other than the tōjisha themselves. While expanding to the field of disability studies, the term gradually shifted away from the legal definition of a "concerned party" toward "a subject position (*shutaisha*) based on shared characteristics" and began to refer to "individuals with direct experience of discrimination" (196). As a result, the term came to be associated with the field of disability studies especially under the themes of discrimination and human rights.

It is important to point out that, unlike physical disability, intellectual disability and mental disability are two areas that have been neglected in the disability rights movement in Japan (Nakanishi and Ueno 2003/2015), although that is not unique to this country. People in the intellectual and mental disability categories are often deemed "incapable" of having good judgment and making proper decisions on their own. Seen as too incapacitated to make choices for themselves, they may be required to have someone else, a representative, decide for them. Involuntary hospitalization used to be permissible because of their perceived lack of reasoning capacity, and tōjisha were often hospitalized against their will. This trend still continues, Nakanishi and Ueno argue, and of all the developed nations in the world, the Japanese medical system currently possesses the largest number of hospital beds for the mentally ill. For those with mental disabilities, however, a breakthrough occurred recently: a new wave of tōjisha movement has arisen in a group home called Bethel House in Urakawa City, Hokkaido. More information on Bethel House can be found in the section on *tōjisha kenkyū*.

Pressing with *Tōjisha Sovereignty*, people with disabilities succeeded in separating their needs from what medical professionals demand or desire (McLelland 2009), fight for their needs and rights under a single banner (Teruyama 2011), and address each other respectfully within their own community using the term tōjisha (Nakamura 2013). In a larger context, the term tōjisha is not just associated with disability or feminism; however, in the academic literature, the term is used to refer to those who are in traditionally stigmatized predicaments such as sexual minorities (e.g., McLelland 2009), victims of domestic violence (e.g., Shinoda 2001), and by far, in recent publications on neurological differences such as Asperger's syndrome (e.g., Ayaya and Kumagaya 2008; Teruyama 2011) and mental health issues (e.g., Bethel House 2005). Kodaira and Ito 2011; Kumagaya 2011; Kim 2013; Matsuda 2018; Sato 2015).

Caregiver/Family Tōjisha

Here, a critical question is whether a primary caregiver in the family can possess *tōjisha sei* (tōjisha-ness) and thus can be a tōjisha. It is clear that tōjisha means a person with a mental or physical disability in the context of disability studies. In their lives, a primary caregiver family member (e.g., spouse, adult child) plays a pivotal role in their welfare, particularly if they are a child or adult who is unable to take care of themselves especially in the midst of their mental health crisis. In this regard, the caregiver shares the lived experience of mental health challenges with the tōjisha. Sato (2015) emphasizes the fact that being the closest caregivers, family members shoulder the enormous responsibility of ensuring the disabled person's welfare (which society also takes for granted) and can easily drive themselves into the ground by not sufficiently resting and prioritizing the needs of the cared-for over their own. In this tight-knit interdependence between the two parties, she argues, even if the caregiver has no disability, they do develop shared tōjisha-ness. Furthermore, as I cited earlier, Nakamura (2013) also translates the word tōjisha as "people with disabilities or families thereof" (247).

Of course, we must not ignore the fact that the tōjisha's needs and their families' needs do not always match (Nakanishi and Ueno: 92–93). For example, it is very common that the tōjisha feel powerless because they are isolated and their voices are not heard within their own communities. Furthermore, the family's voice is almost always given more weight than that of the person with the disability. Typically, parents have more say in decision-making, especially when the tōjisha are minors. Under Japan's family structure, childcare is traditionally considered to be the parents' responsibility because they gave birth to their children. If a child is born with a disability, then the parents—primarily, the mother—are blamed for that "problem," and they are obliged to do everything they can for the child at any costs. With such a heavy burden, there have been multiple cases in Japan in the past in which the mother killed the disabled child and herself in *michizure shinjū*, or double suicide (Nakanishi and Ueno: 150–151). Sato (2015) herself reports an incident that occurred on May 28, 2015, in which a mother killed an intellectually disabled daughter in Hokkaido. In this sociocultural background, it is of primary importance to ensure the priority of the needs of those who are cared for rather than those of their caregivers (as long as the situation does not lead to caregiver exhaustion and the danger to both parties' health).

I was initially ambivalent about the application of the word tōjisha to the caregiver. As Karen Nakamura suggested to me in our panel discussion at the 2019 Association for Asian Studies Conference, the two parties are distinguishable as *tōjisha honnin* (the very person with a disability) versus *kazoku tōjisha* (family tōjisha). Indeed, Nakamura's point has a practical value in clarifying whose voice is being represented in the story. At the same time, many of us *kazoku tōjisha* such as Carolyn Stevens (also on our panel) underscore an inseparable, symbiotic relationship between the two. Stevens, author of *Disability in Japan*, commented on my presentation on this subject by insisting that caregivers like us are legitimate tōjisha as *kaigo tōjisha* (caregiver tōjisha). We family/caregiver tōjisha are survivors of relentless challenges and possess tōjisha-ness as the result of sharing the lived experience of ostracization, discrimination, and ceaseless challenges. Who would say that our voices are unimportant? It is this perspective that encouraged me to include manga by family caregivers in this book. Consequently, two of my analysis chapters are on *tōjisha manga* stories written and illustrated by caregiver tōjisha.[1]

Significance of Lived Experience

Emphasis placed on tōjisha voices relates to the disability rights motto, *Nothing about us without us.* The motto has attracted attention from disability studies scholars since the publication of *Nothing About Us Without Us* by James Charlton (2000). As he points out, oppression experienced by people with disabilities is akin to the oppression of racism, sexism, and colonialism. His findings from interviewing individuals with various disabilities for more than ten years attest that the tōjisha knows what is best for themselves and thus should be given top priority in decision making. In a Japanese booklet published by the Japanese Council on Disability (*Nihon Shōgaisha Kyōgikai*), the author Fujii (2014/2016), a blind man and the president of that council, explains that the phrase *Nothing About Us Without Us* was derived from the International Covenant on Economic, Social and Cultural Rights, a treaty adopted by the United Nations General Assembly in 1966 (14). Fujii has lived experience of depression and understands the difficulty of promoting awareness for such "invisible" disabilities. Disability memoirs are critical as we are able to access the tōjisha's perspectives directly. Chapter "Tōjisha Narratives" of this book expands upon this view and explores the critical role of memoirs, particularly graphic memoirs, in contrast to comics about mental health issues written by non-tōjisha.

The following reasons also account for why I believe (1) that it is important to hear tōjisha voices; (2) that academic and medical interpretations of disability alone are insufficient; and (3) that tōjisha perspectives must be integrated into all forms of mass media, including manga. First, ignoring their voices and not reflecting their experiences will continue to perpetuate stereotypes and inaccurate messages about mental health problems. Hearing directly from the tōjisha will help us recognize biases and misunderstandings that persistently arise in public discourse as well as condemn any act of stigmatizing and "othering" people with mental conditions. Second, the tōjisha's needs and concerns must be heard and their voice must be amplified for disability advocacy. Unless the tōjisha stand at the front line in the creation of disability narratives as did these manga artists I interviewed, we will continue to let myths, stereotypes, and ignorance about individuals with disabilities dominate our public discourse. As emphasized in the previous section, the idea behind the term tōjisha is the indispensable value of the lived experience of people with mental health issues. Thus, I argue that foregrounding their voices in publications, particularly popular culture media that is perceived as most accessible, is of utmost importance if we truly desire to create a society where those with and without lived experience coexist in mutual respect.

Tōjisha Undō

Two concepts emphasized in the aforementioned discussion of lived experience of mental disability are *tōjisha shuken* (*Tōjisha Sovereignty*, and *tōjisha mesen* (tōjisha perspective). These are highlights of *tōjisha undō*. Technically speaking, *tōjisha undō* refers to Japan's minority movement, in general. However, framed in the context of disability studies, it is also translated as Japan's disability movement (e.g., Ishihara 2015; Tateiwa and Sugita 2016).[2] There was a time when tōjisha voices were not taken seriously in the academic world, but because of the increased number of narratives written by the tōjisha, their perspectives were brought into the spotlight as *tōjisha kenkyū* (Matsuda 2018). The next section will define this key term, *tōjisha kenkyū* (self-study). For now, let us focus on what the tōjisha movement tells us about Japan's attitudes toward mental disability in particular.

As mentioned earlier, disability is conceptualized in three groups in Japan: physical disability, intellectual disability, and mental disability. Matsuda (2018) provides a brief history of the tōjisha movement specifically related to the third group. Since the 1960s, life stories by authors with a mental disability have been published in Japan. The first support group for people with a mental disability, *zenkoku "seishinbyō sha" shūdan* (National "Mental Patients" Association), was established in 1974. Although this organization was originally designed to advocate for patient perspectives, their narratives fell short of entering the academic field as a research area. A 2009 book, *Shuki kara manabu tōgōshicchōshō*, documents tōjisha's lived experience with schizophrenia published from 1966 to 2007 in Japan. It was not until the 2000s that scholars began to listen to and document tōjisha's life stories. At around the same time (shortly before and after 2000), Japan also aimed at dissolving long-term hospitalization and transferring patients back to their communities.

Peer-support-group organizations such as the aforementioned National "Mental Patients" Association played the key role in that transition. In 2006, *Shogai-sha Jiritsu Shien Hō* (the Act for Supporting the Independence of Persons with Disabilities) was enacted.[3] That law was severely criticized and was revised as *Shōgai-sha Sōgō Shien Hō* (the Act for Supporting All Aspects of Persons with Disabilities) in 2013. Matsuda claims that, to this day, this decree has been "a very effective law to support ex-patients' acclimation into their local communities" (153).

Recall that the Ministry of Labor and Welfare's year 2004 goals were not only to improve the public's understanding of mental disabilities but also to shift from hospitalization-centered psychiatric care to more community-based living (Kodaira and Ito 2011). In 2014, the ministry published its review committee's summary of concrete measures that aim at Japan's psychiatric hospital reform via their long-overdue goals of "hospital bed reduction" and "shift from long-term hospitalization to community livelihood of mental patients" (Matsuda 2018).

Western Equivalents of Tōjisha and the Tōjisha Movement

As mentioned in the introductory part of this chapter, the term *psychiatric survivors* is roughly a Western equivalent of tōjisha and is one of many self-referential labels widely used by mad studies scholars and activists, including *mental health consumers*, *psychiatric system survivors*, and *ex-patients*.[4] Considering that *tōjisha undō* speaks to the importance of patients' voices in research on mental disabilities, Japan's movement for and by tōjisha is akin in many ways to the mad movement that began in the early 1980s in the West. To provide a broader context to situate this chapter's arguments for tōjisha, I will now introduce Anglophone points of view on mental disability, particularly through the scholarship of mad studies addressed in English-language publications. Thus, this section will shed light on Western concepts associated with mental disability such as "psychiatric survivors" and "madness" advocated in mad studies.

In the scholarship of mad studies, the term *mad* (or *madness*) is used in reference to psychiatric survivors' mental health conditions, as opposed to medicalized terms such as *mental illness*. This term began to be applied as a way of self-identifying with the mental patient liberation movements in the 1960s and is now used as an "umbrella term" to represent a diversity of these identities associated with mental health conditions (Lewis 2009).[5] Connecting academic critiques and the Mad Pride movement[6] of activists, artists, and intellectuals, the term has been adopted to challenge the normal-abnormal binary and doctor-patient hierarchical attitudes of mainstream psychiatry (Lewis 2009). Similar to *tōjisha undō* in Japan, the significance of the lived experience of psychiatric survivors is the central issue of the Mad Pride movement (simply "the mad movement" hereafter).

Out of the mad movement, which had been active in Vancouver and Toronto for twenty years and in other parts of the world, a new discipline sprung (Ingram 2016). Named "mad studies," the discipline is a field of disability research and activism led by psychiatric survivors. It first emerged in England and Canada and then subsequently developed further in the United States. The very first writing that combined all matters affecting psychiatric survivors within a Canadian context was Mel Starkman's essay published in 1981 in *Phoenix Rising*, a psychiatric survivor magazine published from 1980 to 1990 (Menzies et al. 2013). Starkman, a Toronto-based psychiatric survivor and activist, wrote, "An important new movement is sweeping through the Western world ... The 'mad,' the oppressed, the ex-inmates of society's asylums are coming together and speaking for themselves" (cited in Menzies et al.: 12). Richard Ingram, another key figure in psychiatric-survivor activism in Canada, coined the term *mad studies* at the Madness, Citizenship and Social Justice conference at Simon Fraser University in Vancouver in 2008 (Gorman and LeFrançois 2019).[7]

As the study of madness and psychiatrization (process of being psychiatrically treated), mad studies (also spelled in the upper case as "Mad Studies") is concerned with the oppression and agency of psychiatric survivors (also "Mad subjects"). This new discipline is the first systematic critique of psychiatry that placed psychiatric survivors at the center of scholarship and incorporated critical race and post-colonial understandings.[8] In his subsequent paper, Ingram (2016) reveals that he was "locked up in the summer of 2008" in a form of "psychiatric confinement" (12), the same year he was the keynote speaker at that conference. The studies document psychiatric survivors' fights against the institutions of psychiatry and challenges mainstream psychiatric discourse in the academic realm. Ingram (2016) states that patients' voices have been suppressed ever since "the very beginning of organized psychiatry in Canada and abroad" and that the field of mad studies expresses "long-standing psychiatric survivor narratives" with a "burgeoning form of activism"; therefore, this activism is "an alternative to Big Pharma[9] and profiteers in the psychiatric system" as well as academics "who make a living labelling and medicating" psychiatric survivors like him (2). Inheriting the spirit of the antipsychiatry movement, mad studies is connected to human rights movements such as women's rights and disability rights, among others, and contributes to the understanding of the nature of psychiatric conditions and practices (3). As we have seen, mad studies scholars value lived experience of people with a mental disability and advocate psychiatry-survivor analysis as legitimate research methodology, just as the significance of tōjisha voices and of life writing by the tōjisha is beginning to be recognized in Japanese academic institutions. In this regard, the mad movement is the Western equivalent of the tōjisha movement.

Activists and scholars in disability studies have challenged the dominant social value of "normality" by arguing that acknowledging all types of embodied human differences enables us to eliminate, or at least reduce significantly, the stigma and prejudice attached to the medical model of "impairment." However, British scholars McWade et al. (2015) point out that the field of disability studies has failed to include perspectives of people with lived experience of a mental or neurological condition. McWade et al. also contend that even those academics who align themselves with the psychiatric-survivor movement continue to frame patients/survivors as outsiders, particularly if the academics themselves are not psychiatric survivors, and treat the survivors as subjects to research "on" rather than informers to work "with." While pointing out these shortcomings of North American disability studies scholarship, they urge us to focus on individual suffering and move toward a collective identity, an argument that seems to resonate with the tōjisha arguments raised in Nakanishi and Ueno's *Tōjisha Sovereignty* (2003/2015). While addressing inequalities such as disablement, discrimination, and marginalization affecting people with lived experience, McWade et al. (2015) assert that key concepts such as recovery, inclusion, and access that befall people with physical disabilities are equally germane to those affected by mental and neurological disabilities. Therefore, they urge both psychiatric survivors and neurodivergent individuals to bring the shared experiences of oppression and marginalization and run anti-stigma campaigns together with people with physical disabilities.

Similarly, a US disability studies scholar, Margaret Price, argues that the social model hoisted in disability studies fails to acknowledge the fact that certain types of "physical pain and suffering" experienced in mental disability are not necessarily brought by society, unlike the case of physical disability. Although no direct reference to mad studies was made, Price's (2015) *Mad at School* parallels writings of Canadian psychiatric survivors and those of Japanese *tōjisha undō* activists. *Mad at School* is considered a must-read on the topic of mental disability in higher education by many US scholars (e.g., a recommended read in the 2018 National Endowment for the Humanities Summer Institute on disability studies). In the book, Price (2015) adapted the term *bodymind* from trauma studies while reflecting on her lived experience of a mental health crisis. She uses the term to call for the close connection between our body and mind in academic discourses on mental disability, stressing the need for more attention to pain in disability studies. Price tells of her own experience of mental breakdowns while investigating how psychological illness affects academic culture and challenging the social model approach to understanding "madness" from her disability-studies perspective.

Although I described the social model of disability in detail in *Reframing Disability in Manga* (2020), I am acutely aware that simply advocating this model is insufficient when it comes to raising awareness of first-person accounts of mental health experiences. To augment, I will now highlight *tōjisha kenkyū*, a particular field of study that integrates tōjisha perspectives into academic research about psychological or neurological conditions. Examining the field of *tōjisha kenkyū* will also help decentralize Euro-American centrism in disability studies. That way, I hope that the field of disability studies continues to benefit from non-Western examinations of disability and stigma.

Tōjisha Kenkyū

In comparison to mad studies in the West, Japan's *tōjisha kenkyū* have a more symbiotic, less contentious relationship with mental healthcare providers. This section examines what *tōjisha kenkyū* means while highlighting similar and different beliefs about mental health service users between these two schools of thought.

No agreed-upon translation of *tōjisha kenkyū* exists in English partly because it is still a relatively unknown field of study to the English-language scholarship. For example, Kumagaya (2011) translates the term as "person-centered, peer-supported research," and Ishihara (2013) defines *tōjisha kenkyū* as "research" to be conducted by the tōjisha, who have a disability or problem themselves, by confronting their own issues and studying them with peers. Simple English-language equivalents such as "self-research" or "self-studies" are also employed (e.g., Kaneko 2019). Explaining more elaborately in her book, Nakamura (2013) calls *tōjisha kenkyū* "research on yourself-as-part-of-an-affected-class" and uses the English translation of "self-directed research" or "self-directed research projects," although she warns that either translation "unfortunately negates some of the political aspects of the term *tōjisha*" (173). To be sure, any condition can be studied as *tōjisha kenkyū*. However, it is important to point out that *tōjisha kenkyū* was originally begun among residents of Bethel House who had schizophrenia and has been further developed by tōjisha with *hattatsu shōgai* (developmental disabilities) or those with lived experience of mental health problems. Furthermore, usually as a self-directed research project conducted for and by an individual with a stigmatized mental health condition, *tōjisha kenkyū* has become an indispensable component of legitimizing tōjisha's lived experience scientifically in Japanese public discourse (Ishihara 2015).

Ishihara (2015) explains what makes this approach a *kenkyū* (study/research): similar to scientific research, it consists of information gathering and hypothesis testing through trials and allows the person who conducts it to "fail to realize the expected result" (30). In addition, its main function is "to give back to individuals with mental health problems the right to appear and talk in the public sphere" (30). This research enables the tōjisha's self-redefinition and reconnection to others in a healthy way through *tōjisha gatari* (*tōjisha* storytelling) while working in tandem with Social Skills Training (SST) and the techniques of peer-support groups (Ishihara 2013: 12). The research also entails *ninchi kōdō ryōhō*, or cognitive-behavioral therapy (CBT). Mukaiyachi has published a book (Ito and Mukaiyachi 2007) about Bethel-style CBT, in which CBT is defined as a self-help tool that allows the tōjisha to perceive the mechanism of their own troubles and find an escape route. In Bethel-style CBT, experts are there only to assist the tōjisha in the process. Unlike the self-psychoanalysis of looking deep into oneself, this method focuses on two goals: identifying ways in which they understand issues surrounding them and acting upon the realization in deciding how to deal with the issues themselves. Bethel's *tōjisha kenkyū* and SST in particular are key to approximating the goals as these enable them to explore the universal and common aspects of lived experience of psychological distress with other human beings (Ishihara 2013). Nakamura (2013) offers a benefit analysis, arguing that the formal, scientific connotation of *kenkyū* helps the tōjisha gain a psychological distance from and more objective attitudes toward the studied themselves when observing their own symptoms. In these regards, *tōjisha kenkyū* is not an equivalent of diary writing but rather a form of research projects for and by the tōjisha.

Historically, *tōjisha kenkyū* was originally developed as a tool of empowerment for people with mental health problems. Prior to its emergence, some sociocultural conditions existed. First, as mentioned earlier, in the late 1970s in Japan, the previously dominant "third-party perspective" (*daishansha mesen*) began to be criticized in the discipline of sociology, and researchers found fault with the absence of subjects' voices in publications. That insight from sociology was adopted in the fields of anthropology and ethnology, and ethnography, participatory observation, and life-story/life writing were developed as new research methodologies in the 1980s. Second, by the 1990s, specialist knowledge and its political connection were scrutinized while the representation of *tōjisha-sei* (tōjisha-ness) was emphasized. Third, in the 2000s, *tōjisha-gaku* (tōjisha studies) became a hotly debated topic of academic discourse as a fresh insight into various disciplines, including disability studies. Exemplified by Nakanishi and Ueno's (2003/2015) work, *tōjisha-gaku* has now established itself as an academic field that collects and analyzes tōjisha's subjective experiences, separating them from the objective, outsider perspectives of *hi-tōjisha* academics and medical professionals.

It was in this sociocultural backdrop that *tōjisha kenkyū* was born in the early 2000s. Its birthplace was a group home for people with mental health issues located in the small fishing town of Urakawa in Hokkaido, Japan (Nakamura 2013; Mukaiyachi 2015; Ishihara 2015). The group home is called Bethel House (*Beteru no ie*).[10] The practice of *tōjisha kenkyū* began by chance in February 2001, about two years before the original publication of Nakanishi and Ueno's book, *Tōjisha Sovereignty* (Ishihara 2015). That year, one of the residents of Bethel House, Kawasaki Hiroshi, agreed to engage in a *jiko kenkyū* (self-study) suggested by Mukaiyachi Ikuyoshi.[11] As Mukaiyachi (2015) himself recalls in the introduction to his 2005 volume, the first project of *tōjisha kenkyū* came into existence in 2002 (a slightly different history from Ishihara's account) while he was working with Kawasaki who had been hospitalized with schizophrenia. One day, Kawasaki "blew up," one of his symptoms. Mukaiyachi then asked him, "What would you say about doing research together on how to get along with the person named Kawasaki Hiroshi?" Mukaiyachi saw the sparkle in Hiroshi's eyes. Kawasaki responded immediately, "I'll do it." The word *research* attracted Hiroshi and helped him see the possibility of overcoming his feelings of helplessness.

The practice of self-study (then called "*jiko kenkyū*") soon spread among members of Bethel House (Ishihara 2013: 30). That is because soon after establishing the business of seaweed packaging,[12] the members of Bethel House expanded their products to various commodities and integrated the approach of *kenkyū* (research) into their business practice for discussing sales strategies and brainstorming the development of new products (Ishihara 2015). Once their attitudes toward "research" were well established, the members adopted the same approach to examining their own mental health problems, naming it *tōjisha kenkyū*. Through this approach, the Bethel House members were united with the idea of working on their personal issues together as peers based on their philosophy that they have the ability to recover relationships with other human beings and can support each other. *Tōjisha kenkyū* then gradually spread among individuals with mental health issues beyond the little town of Urakawa. By the early 2000s, it was "rapidly becoming an acclaimed aspect of treatment at Bethel" and is now "a hallmark of Bethel presentations all across the country" (Nakamura 2013: 172). Along with publications by Bethel House and Mukaiyachi, works such as Satsuki Ayaya and Shinichiro Kumagaya's studies also contributed to the recent development of *tōjisha kenkyū* (Ishihara 2015: 30).

As Kodaira and Ito (2011) explain, Bethel House has thrived through various innovative events (e.g., the annual Bethel festival, "Illusion-Delusion Convention") and unconventional ways of approaching mental disability (e.g., upon meeting with one of the most troubled schizophrenia patients, Mukaiyachi introduced himself as a social worker who came to "seek advice" from him). In this environment, the tōjisha begin to reflect upon their personal growth and life philosophy while battling their own mental health issues. Bethel principles emphasize that each tōjisha is a member of the community like anyone else so that they develop a sense of belonging and real-life social participation. Many principles under which Bethel operates are derived from either Christian philosophy or Alcoholics Anonymous guidelines. The core principle is *sando no meshi yori miitengu* (literally, "meetings outvalue three meals of the day"). Run by Bethel members themselves, meetings in which participants tell each other how they are doing both physically and mentally, how *kombu* seaweed sales went, and so on take place constantly every day. Bethel has flourished also because of a good partnership between the social worker

Mukaiyachi and Dr. Kawamura, the chief psychiatrist of the Urakawa Red Cross Hospital (Mukaiyachi 2015). Dr. Kawamura describes the group home as a groundbreaking culture that has never before been found in the mainstream medical world.

The unique, beneficial characteristics of *tōjisha kenkyū* have been recognized by scholars in Japan (e.g., Nakanishi and Ueno 2003/2015; Ishihara 2015). The key tenants of this tōjisha-led self-study are (1) the tōjisha can "rest on the shelf" while facing their problems; (2) they recover their right to "talk about their problems in the public sphere as responsible agents by addressing them in the form of a study"; (3) they recover their "personal histories" and reidentify them; (4) they can "enter into a study community and enhance connections with their peers" (Ishihara 2013: 64–66 in Japanese; Ishihara 2015: 34 in English). These principles have been publicized through publications, including a collection of *tōjisha kenkyū* reports, *Bethel House's Self-Directed Research* (Mukaiyachi 2015) and two books under the trademark of Bethel House, *Let's Do Self-Directed Research* Vol. 1 (Bethel House 2009) and Vol. 2 (Bethel House 2011).

Prior to the 2000s in Japan, being admitted to a mental hospital was deemed the "most miserable" thing in life, and recovering ex-patients saw little or no hope in the community of Urakawa as a place for rehabilitation (*shakai fukki*) (Ishihara 2013). However, *tōjisha kenkyū* arose as a complete game changer. First, it was designed to give the Bethel House members "insight into their illnesses" while providing "a socially sanctioned and ritually defined narrative structure in which to talk about their experiences and feelings" (Nakamura 2013: 172). Second, with one of the goals of *tōjisha kenkyū* being making formal research presentations at various venues, including presentations for pharmaceutical representatives, talks at regional conferences, and the annual Bethel Festival in the summer, the tōjisha would conduct their own *tōjisha kenkyū* and become able to elevate their status from the traditional "research subject" studied by doctors or academics to researchers themselves. This new status helped the tōjisha connect to the community more comfortably and represent their voice confidently in the public discourse about mental health. In other words, it enabled them to openly share the narrative of their lived experience and "access the public sphere" unlike generations ago (Ishihara 2015: 34).

Of the characteristics of *tōjisha kenkyū* described earlier, the tōjisha's willingness to collaborate with pharmaceutical representatives, in particular, stands out when compared to the Western movement led by psychiatric survivors. Unlike psychiatric survivors, whose aim is to redefine the label of "mental illness" by challenging psychiatry and pharmacological treatments, promoting a critical discourse of "madness," and perceiving the world in different ways, the participants of *tōjisha kenkyū* are quite open to working with medical professionals.[13] Kodaira and Ito (2011) argue that the success of Bethel House is a clear indication to the medical field that it is possible to integrate ex-patients into the community and serves as a progressive model of mental disability inclusiveness. Mukaiyachi himself admits that recent years of Bethel House activities are remarkable and have attracted attention from people who engage in various types of caregiving jobs, dubbing its advancement "the era of tōjisha advocates" (Kodaira and Ito 2011).

One reason that *tōjisha kenkyū* differs from mad studies might be how tōjisha is defined at Bethel House. In Urakawa, the stance of tōjisha presented in the study's perspectives originated in Mukaiyachi's philosophy: "I should not make decisions about myself alone" (Mukaiyachi 2006: 68) and "It is about ourselves that it is most difficult to know"[14] (Mukaiyachi 2009: 44) (cited in Ishihara 2015: 35). Mukaiyachi was largely influenced by Viktor Frankl's existential analysis whose chief idea is that we all have the responsibility of choosing a certain type of attitude upon facing the fate that cannot be changed (Ishihara 2013). Based on Frankl's philosophy, Mukaiyachi believes the essence of *tōjisha kenkyū* is assuming the responsibility of living with the fate; even if the tōjisha find themselves in the circumstances beyond their control, they must assume that responsibility willingly. As Ishihara (2015) explains, the primary goal of *tōjisha kenkyū* is to provide the opportunity for the tōjisha to face their own problems and explore them together with peers in a *kenyu* (research)-style. With that being the priority, the participants of *tōjisha kenkyū* do not claim that service and research should be provided or controlled only by the tōjisha and instead believe that "academic knowledge and professional support are not to be excluded from *tōjisha kenkyū* but to be used as resources" (Ishihara 2015: 37). Interestingly, this collaborative attitude toward academia as well as the field of psychiatry is also evident in almost all the graphic memoirs about a psychological or neurological condition, more than fifty manga titles that I investigated in Japan (this point will be further described in the case study chapters).

In this way, *tōjisha kenkyū* also differs significantly from *tōjisha undō*, Japan's disability movement described earlier. For example, while the former presumes that "patients do not know themselves well and need to study themselves" (they need to "study themselves" to become capable of "addressing their own problems"), the movement's motto was the opposite: "we know what we need; we decide what service should be provided" (Ishihara 2015: 27). As I pointed out before, Mukaiyachi's perspective of tōjisha has been greatly influenced by Viktor Frankl's existential analysis. It is with this premise that *tōjisha kenkyū*'s self-study facilitates "the admittance of one's own weakness and providing a public place to discuss that weakness" (27).

To reiterate, Mukaiyachi's view on who should be taken into account as tōjisha sharply contrasts with the definition of Nakanishi and Ueno: "When someone has needs, he or she is already a tōjisha" and "tōjisha refers to the end user of the service" (Nakanishi and Ueno 2003/2015: 4). In Nakanishi and Ueno's view, there is no hierarchy like the more severely disabled one is first considered as tōjisha. Their idea of tōjisha seems to be based on the idea of service-user rights. Thus, they say, "Tōjisha know about themselves best" (Nakanishi and Ueno 2003/2015: 12). Their model of *tōjisha shuken* (tōjisha's sovereignty) is based on the premise that "I am sovereign of myself, nobody except me—state, family, professionals—is allowed to decide who I am, what I need on my behalf" (Nakanishi and Ueno 2003/2015: 4)—similar to James Charlton's (2000) "Nothing about us without us" in the West (Ishihara: 35). Although this representation of tōjisha poses a huge contrast between *tōjisha kenkyū* and mad studies, Ishihara argues that it is because *tōjisha kenkyū* places the priority on "the alteration of the patients' attitudes toward their suffering, facilitated by their taking part in a study community" (Ishihara 2015: 36).

As shown earlier, *tōjisha kenkyū* emphasizes the modification of the tōjisha's attitudes "toward their problems" (Ishihara 2015: 34). In other words, the tōjisha is represented in a uniquely different way from the philosophy of the service-user rights model from which the psychiatry-survivors movement developed. Another key difference is that *tōjisha kenkyū* is "not necessarily hostile to psychiatry" (34); medical professionals are "authorities on medical diagnosis and medical treatment" (35). That attitude is reflected in one of Bethel House's activities in which the tōjisha create names for their own conditions, often including parodies of the diagnosis used in the DSM. In this naming activity, the tōjisha "are just making use of the established medical labels and do not concern themselves with the hegemony of naming politics" (35). Although they are aware that medical labels such as schizophrenia are insufficient for them to describe the nature of difficulties they experience, they find the medical terminology "useful in naming their conditions" and just adapt them to create their names (35). Thus, in *tōjisha kenkyū*, it is considered unnecessary to stand in for the professionals or control their research.

It is a fact that a gulf of knowledge exists between doctors' and patients' lived experience of mental disability. Thus, collaboration between doctors and their patients is seen as a great advantage in Japan. This gap of experience has frequently been pointed out, and an increased effort has been made to incorporate tōjisha's voices in research in the field of medicine in recent years (Kaneko 2019) as is evident in the case of *tōjisha kenkyū*. For this reason, it is considered a benefit to psychiatrists to receive the tōjisha's input, particularly on not-so-easily-observable symptoms, because such insider information can lead to the development of a new treatment.

Furthermore, the principle of "self-discovery" together "with peers" propagated in *tōjisha kenkyū* has tremendous merits as the tōjisha tend to reveal "their own weaknesses" and "their experiences of hardship" to the other participants through narratives (Ito 2011). Quite often, psychiatric symptoms are hard to describe precisely. Thus, during the meeting at Bethel House, the tōjisha attempt to describe their lived experience and life strategies to others by using metaphors, analogies, and personifications of symptoms and problems such as "*genchō-san*" (my auditory hallucination), "*okyaku-san*" (my visitor), "*shintai no gosadō*" (malfunction of body and mind), and "*bakuhatsu*" (blowing up). The naming of a symptom is the ritual of affirming that the person has lived life bravely without giving up as well as a process of self-rediscovery. The step of creating a self-diagnosis instead of simply referring to the given medical diagnosis allows the tōjisha to feel ownership over their illness (Nakamura 2013). In addition, the naming encourages the tōjisha to differentiate between the self and their problems. For example, Hiroshi can introduce himself to his members, saying "I'm Hiroshi who is struggling with the issue of how to stop blowing up (*bakuhatsu suru*) even when I don't want to" instead of "I'm Hiroshi. I have schizophrenia." Such creative and narrative self-diagnosis helps his peers relate to Hiroshi's struggle and feel invited to analyze his issues together (Nakamura 2013). Similarly, this inventive process encourages them to adopt the attitude of using humor while grasping the meaning of their own health problem; thus, they always laugh a lot during their meetings; because humor originally meant "laughing regardless," Mukaiyachi believes laughing is the ultimate form of "courage to live" (*ikiru yūki*) (cited in Ito 2011). (Chapter "Essay Manga" of this book has more discussion on the use of humor in disability memoirs.)

It is important to point out that, besides the importance of humor, the role of peer support is one of the key elements of group meetings in *tōjisha kenkyū*. While the tōjisha are trying to externalize the understanding of their own symptoms through the words and drawings they put on a whiteboard, the other members either observe or help reorder their issues in ways in which the tōjisha separate themselves from their issues. Along with the personification strategy of naming their symptoms, the strategies of externalization and objectification of the problem they engage in with peer assistance enable the tōjisha to deal with their symptoms in more practical manners. Another impactful practice is role-playing: after sharing their narratives, they engage in peer-to-peer discussion and later act upon the suggested solutions for internalization. When each other's stories are shared, a sense of solidarity is formed; thus, even in the stories of personal "failure" or "standstill," one can find important clues for the future and other resources (Ito 2011). Sharing their story of experiencing a serious mental health issue with others can serve as a means for the tōjisha to connect with the outside world and empower themselves in society.

How does this process of presenting one's ongoing issues and receiving peer advice actually work for the tōjisha in their daily meeting? Nakamura (2013) reports as follows based on a meeting she observed at Bethel House: a person diagnosed with a severe anxiety disorder discussed his fear of experiencing a panic attack in a crowd; his peers offered suggestions on how to create a relationship that accepts his panic attacks; and then he felt "accepted" as who he was by his peers and was no longer afraid of having a panic attack. This peer advice system also works to help tōjisha see themselves on equal footing with their medical professional counterpart; it puts them in the position of the authority examining their conditions (Nakamura 2013: 197–200). The importance of this peer advice system is well summarized in the advertising message that appeared in *Beteru-no-ie no "Tōjisha Kenkyū"* (Bethel House 2005): "*Tōjisha kenkyū* means a festival of hoisting your own *mikoshi* of hardship with your peers" (a *mikoshi*, a portable shrine, is a metaphor for something to put on one's shoulders and carry together with a group of people in celebration). Therefore, *tōjisha kenkyū* helps the tōjisha make a community by collaborating with each other (Ishihara 2015). Even Nakanishi and Ueno find this role of peer support as a great benefit of *tōjisha kenkyū* (Nakanishi and Ueno 2003/2015).[15]

The spirit of collaboration emphasized in the philosophy of *tōjisha kenkyū* has turned into the catchphrase of "by oneself, together," and its main purpose is not problem-solving but rather figuring out the patterns of how your problem emerges. Citing the words of Mukaiyachi (2015), Ishihara (2015) stresses that the goal of *tōjisha kenkyū* should be cultivating the attitude of confronting one's own symptoms and resultant social problems by recording how those symptoms occur by themselves. Another important characteristic of this approach is that nobody criticizes the speaker as is evident in its motto, "Let them speak and let them listen."

In sum, *tōjisha kenkyū* began as a method of self-directed research at Bethel House, a group home established in 1984. It was developed by a small group of tōjisha who have blossomed with the support of their social worker, Mukaiyachi Ikuyoshi. Since *tōjisha kenkyū* was born at Bethel House, it has spread throughout Japan and has been applied to various academic topics, including research on Asperger's syndrome. *Tōjisha kenkyū* has been praised as an activity that no other psychiatric approach has ever succeeded in challenging "the common-sense understanding of mental illness in Japan" (Ishihara 2015: 29), especially in its emphasis on the tōjisha's collaboration "with their peers and often with the help of professional supporters" (27).[16] This approach has been adopted far and wide after a little more than ten years since its beginning. One reason of its successful publicity is that the Bethel members have traveled to many places to advocate and demonstrate the *tōjisha kenkyū* method. Thanks to their efforts, *tōjisha kenkyū* is now fairly well known in foreign countries. For example, it has been popularly applied to research in South Korea since the motto "Nothing about us without us" was introduced into the country where the voice of mentally ill tōjisha had long been suppressed (Kim 2013). Currently, there are about ten group homes in Urakawa. While engaging in *tōjisha kenkyū*, Bethel House members sell local goods such as *kombu* seaweed to make a living; some of the members write books and travel to give talks throughout Japan (Nakamura 2013). It is under these merits of self-discovery, patient-doctor collaboration, and public sharing of one's lived experience that the book title, *tōjisha manga*, resonates and is legitimized to refer to Japan's graphic memoirs about mental or neurological predicaments.

Conclusion

In Japan, it is very common for psychiatrists to spend little or no time with their patients and simply dispense medication without listening to their patients' stories. However, it is critically important to understand the individuals' points of view and their unique experiences in mental health care as they are quite often misunderstood by the general public as having "personality flaws" or of using their condition as an "excuse" to get what they want. Fortunately, in Bethel's daily meetings, the tōjisha discuss delusional voices and visions or other troublesome symptoms with tōjisha peers and staff members in trying to find ways to come to terms with their mental health challenges. Because the main purpose of *tōjisha kenkyū* is for the tōjisha to figure out what is best for themselves, it is believed antithetical to the traditional pattern of removing patients from the decision-making process.[17] Thus, in this approach, tōjisha perspectives (*tōjisha mesen*) is of primary importance. Furthermore, by sharing their stories of lived experiences with the public, they deny themselves the act of seeing their situation as a "personal failure," unlike their predecessors.

As I defined in the Introduction, *tōjisha manga* are real-life stories that narrate the tōjisha's experience of mental health problems. Tōjisha has been adapted as the thematic word of this book to refer to people with lived experience of mental health issues and similarly stigmatized medical conditions because, as I have argued, this emerging group of graphic memoirs deserves scholarly attention without an unsavory label of *seishin shōgaisha*. Arising against the historical and sociocultural backdrops of tōjisha described in this chapter are eye-opening, serious-yet-humorous manga titles narrated by tōjisha or family/caregiver tōjisha who have experienced mental health challenges while dealing with emotional and physical pain, medication problems, family disputes, and societal discrimination and misunderstanding.

In the past, whether it be a mental health challenge such as depression or a neurological issue such as Asperger's syndrome, the stakes were so high due to strong stigma attached to the condition that tōjisha chose to remain "invisible" rather than reveal it.[18] Therefore, such a condition was rarely self-disclosed in a form of comics in Japan. For this reason, as I argued in the Introduction, we need a non-stigmatizing label—*tōjisha manga*—that draws on the sociocultural status of tōjisha described in this chapter to clearly distinguish this emerging group of graphic memoirs about the author's lived experience from other, general autobiographical comics of Japan. The next two chapters present arguments for why the genre of memoirs, particularly graphic memoirs, of tōjisha deserves serious scholarly attention.

Notes

1. As exemplified by one of the case studies in this book, living closely with tōjisha can also lead to a serious health issue in the caregiver. Similar to other developed countries, the burden of caregiving is imposed on female family members—wives, mothers, and daughters—in Japan. Even if the caregiver is spared a serious health consequence, it can be a lonely, tough battle for wives to meet the constant demands of managing a mental health crisis as shown in two case studies in this book. A disability studies scholar, Michael Rembis, pointed out that such "relentless challenges that female caregivers face are brought on not only by the tōjisha but also by society's expectations and how that society is structured both physically and culturally" (personal communication).
2. The movement was brought up by disability rights activists after two key incidents occurred in the 1970s (Nakanishi & Ueno Nakanishi and Ueno 2003/2015: 25–7). The first incident involved a mother who killed her two-year-old daughter who had cerebral palsy, and the Blue Lawn Grass Club protested the public petition for reduced punishment. The second one was a conflict that took place in the Fuchu Care and Education Center where the residents protested against maltreatment and oppressive administration. Those who spearheaded the disability movement throughout the 1970s and 1980s were people with cerebral palsy in Japan who were against the government's plan to allow the intentional abortion of disabled fetuses by amending the Eugenic Protection Act (see more about this history in Tateiwa and Sugita 2016). They also helped establish the first Center for Independent Living (CIL) in Tokyo in 1986, which spread throughout Japan in the 1980s and 1900s (Ishihara 2015: 27–28).
3. In the history of *tōjisha undō*, 2006 is a year that marks significant developments. In the same year, *Beteru-no-ie wa itsumo papipupepo*, an illustrated essay about Bethel House, was published, and Ueno Chizuko's writings about tōjisha were also in print.
4. *Mad Matters* (Menzies et al. 2013) is a book published in Canada to assemble scholarly writings of the mad movement. Also recommended as further reading is *Disability Incarcerated: Imprisonment and Disability in the United States and Canada* (Ben-Moshe et al. 2014), a collection of papers written by Canadian and US scholars of mad studies. The essays in *Disability Incarcerated* will help the reader understand that people with disabilities have been "policed" in complex ways and that a disproportionate percentage of the disabled population has been institutionalized or incarcerated in North America.
5. Put in a sociohistorical context, the term *madness* is evocative of Michel Foucault's *Madness and Civilization: History of Insanity in the Age of*

Reason (1961), and those involved in the Mad Pride movement followed in the footsteps of Black Pride and Gay Pride (Lewis 2009). The antipsychiatry movement is a movement of the 1960s, originally led by psychiatrists and intellectual scholars as a reform movement that challenged the legitimacy of standard psychiatric practice, asserting that psychiatric treatment is not helping but rather damaging patients (Gorman and LeFrançois 2019). In this view, psychiatry is an instrument of patient coercion and oppression via harmful treatments such as lobotomy and is demonstrative of an unequal power relationship between the doctor and patient. By contrast, the current mad movement, which is central to mad studies, is for and by psychiatric survivors; it is a patient-based consumer movement of mental health services that sprung out of the civil rights movements in the 1960s and 1970s. This movement advocates interventions by psychiatric survivors themselves, and its discourse centers around ex-patients' personal histories of psychiatric abuse.

6. Mad Pride events take place in North America as well as in Australia and England (Price 2011).
7. In Ingram's conceptualization, mad studies is "a space of social action and theorizing about oppression and psy-violence that centres the histories of psychiatrised bodies," and its focus is on the "retrieving, documenting, understanding, revisiting and teaching of mad people's history" (Gorman & LeFrançois Gorman and LeFrançois 2019: 107). Although Ingram admits having coined the term, he adds that it was already emerging as a discipline at the time of the 2008 conference at Syracuse University in which he gave that memorable talk.
8. In this discipline's framework, which is based on critical race and feminist theories as well as the histories of colonization, racism, and exploitation of human rights, being psychiatrized is seen as akin to being colonized because patients are seen to be helpless "others" who need to be helped by "superior" medical professionals. By connecting to colonialization, mad studies scholars allude to drapetomania, a discredited category of mental illness hypothesized to cause enslaved Africans to flee captivity, as an example of psychiatrization.
9. The term *Big Pharma* that appeared in Ingram's (2016) writing is a reference to the pharmaceutical industry on a global scale and implies a disparaging view of the industry.
10. The group home was founded by a Japanese Christian group in 1984 (Nakamura 2013; Mukaiyachi 2013). The name Bethel derives from *Beth El* ("house of God") in the Old Testament and is also a reference to the Bethel Foundation, a psychiatric hospital in Germany (Nakamura 2013; Ishihara 2015). The Bethel Foundation is known to have courageously resisted Hitler's orders to euthanize its mentally ill patients in 1940

(Nakamura 2013: 105). Sasaki Minoru, who was discharged from the hospital after having his schizophrenia symptoms treated, moved into one of the rooms at an old church and became the head of the Acorn Society. Then, Hayasaka Kiyoshi moved in, and they renovated the church to make it more habitable in 1984. Reverend Miyajima christened an old church building as "Bethel House." The residents of Bethel House began to package and sell Hidaka ***kombu*** (Hidaka Seaweed) in 1988 and then established the Welfare Shop Bethel in 1989, which began to sell paper diapers and other medical products for the elderly in the community and the patients at the hospital. They also formed a support group, nicknamed "the Acorn Society" (***Donguri no Kai***), for both psychiatric patients and alcoholics and offered meetings in the church once a month. In 1990, Bethel House was renovated yet again for the occupancy of six residents, and the church began to operate three activities: caring for residents at the Group Home Bethel House, packaging ***kombu***, and running a diaper delivery service at the Welfare Shop Bethel.

11. A devout Christian from northern Honshu, Mukaiyachi established a support group for those discharged from the psychiatric ward of the Red Cross Hospital, the region's only general hospital in Urakawa, in the 1980s. He had majored in social work at Hokusei Gakuen University in Hokkaido and was hired as a social worker at the hospital. One of his mission assignments was to help improve the social conditions of Urakawa that had "severe problems with alcoholism and domestic violence" (Nakamura 2013: 102). Mukaiyachi worked hard to help address the problems of alcoholism and domestic violence that existed in the community and brought neighborhood children to the church for Sunday school. Many of these kids were from Ainu families "with histories of mental illness or alcoholism" (104). Mukaiyachi has "a unique charisma, especially within the Bethel community" (Nakamura 2013: 112) and writes books and papers on psychiatric treatment and social work as well as on Christian philosophy.
12. Why seaweed? In the region of Hidaka where Urakawa is located, packaging seaweed was a side job taken by many housewives. The region provided 90% of the ***kombu*** in Japan at that time. Thus, it was a reliable business idea.
13. It is important to point out here that because of the influential philosophy of ***tōjisha-gaku***, scholars of disability studies in Japan do advocate the significance of tōjisha voices, similar to their Western counterpart of mad studies. However, the anti-psychiatry movement by mental service users/survivors developed in the 1970s in the West did not garner much enthusiasm among them in Japan (Ishihara 2015: 33).
14. Toyota (1998) states that even the tōjisha themselves find it challenging to understand their circumstances completely. Similarly, Kumagaya

(Kumagaya 2013) illustrates how unreliable so-called needs are. He quotes the original motto of the tōjisha movement led by people with cerebral palsy: "It is we who know about ourselves best. We decide who we are, what we will do" (218). However, he now questions it and wonders if we always know what we need and who we are. He then proposes to amend the original strong view to a new perspective: "We don't know ourselves well. We ask ourselves who we are, what we will do together with peers" (219). Unless one has a clear, predictable condition, the tōjisha's well-being can change abruptly as their condition may deteriorate due to some factors such as aging. Having experienced the sudden presence of painful symptoms that turned into his *niji-shōgai* (secondary disability), he now knows that physically disabled people, particularly those with cerebral palsy, may sustain a secondary disability as they age. Through this experience, Kumagaya argues that the so-called needs are not stable, nor are they always clear to tōjisha themselves, as he was forced to figure out what his new needs were when his condition suddenly changed.

15. Nakanishi and Ueno (2003/2015) define "tōjisha" as the persons who know themselves well and aim the purpose of their *tōjisha-gaku* at enabling them to reclaim their agency. By contrast, Mukaiyachi (cited in Ogura's paper in Ogura 2014) defines the tōjisha as "persons who least understand themselves," as explained earlier.
16. The growing popularity of *tōjisha kenkyū* does not mean there are no critics. In fact, Yoshii (2014) sorts out a few shortfalls of this approach. First, the goals of *tōjisha kenkyū* are somewhat unclear. Second, unlike the disability rights movement spearheaded by the Association of Green Grass, *tōjisha kenkyū* has no definitive "enemy" to fight against; thus, the tōjisha are not deemed powerful enough to carry the torch for the next generation of the tōjisha movement. Tōjisha share their similar experiences of *ikizurasa* (difficulty of living) through storytelling, but time will test how well such "sharedness" will unite people together. Third, even if their narrative takes the most polished form of storytelling, recipients may only be moved. Its influence falls short of overturning the system of discrimination. Thus, Yoshii criticizes that performers of *tōjisha kenkyū* are hardly aware that they need "weapons" to revolutionalize the status quo. In Yoshii's opinion, unless tōjisha cultivate themselves as the resilient agent of the movement, they will not be able to measure up to the circumstances that discriminate against them.
17. The methods of psychotherapy similar to the philosophy of *tōjisha kenkyū* exist in Japan and beyond. For instance, the method of participant-centric initiatives (PCI) emphasizes the importance of integrating patients' presence, requiring both patients' and their family members' participation in the treatment (Kaneko 2019). Also, the Open Dialogue method in Finland

is a method of intervention, an alternative approach to the treatment of psychosis that started in the 1980s at Keropudas Hospital in Finland (Sato et al. 2017). In this approach, patients have a say about their own treatment in therapeutic "family therapy" meetings and are placed equally to their healthcare providers. Similar to *tōjisha kenkyū*, Open Dialogue utilizes meetings among the patients and peer staff members who assist them. Both Keropudas Hospital and Bethel House are located in a remote area, away from the central cities of each country. In both approaches, the dialogue is structured so naturally that participants are not actually aware that it is a therapy; in fact, in both situations, the approach is considered a "nutrition" rather than a (medical) method. However, they differ in that the former emphasizes equality (equal distribution of decision-making power) and allow a strong teamwork relationship between participants and medical professionals including peer support members, while the latter emphasizes the sharing of participants' self-study themes as well as their pains and suffering with those who are struggling with their own cases of mental illness.

18. Similar to the term *ableism* used in disability studies, *sanism* is a term that refers to an attitude of shaming those deemed "insane," attaching stigma to them, and discriminating and oppressing them just because they have a certain mental trait or condition. However, some may feel ableism is sufficient to critique prejudice against mental disability if we incorporate the view of "bodymind" as discussed in Price's (2011) *Mad at School.* It is beyond this book's capacity to debate whether we need a separate word in discussing stigma and mental disability.

References

Ayaya, Satsuki, and Shin-ichiro Kumagaya. 2008. *Hattatsu shōgai tōjisha kenkyū: yukkuri teinei ni tsunagaritai* (*Developmental disabilities tōjisha research: toward a gradual and respectful relationship*). Tokyo: Igakushoin.

Bethel House. 2005. *Beteru-no-ie no "Tōjisha Kenkyū"* (Bethel House's tōjisha research). Tokyo: Igakushoin.

Charlton, James. 2000. *Nothing About Us Without Us.* Oakland, CA: University of California Press.

Fujii, Katsunori. 2014/2016. *Nothing About Us Without Us: The Trajectory and Nature of Disability Rights Treaty* (*Watashitachi wo nukini watashitachi no koto wo kimenai de: Shōgaisha kenrijoyaku no kiseki to honshitsu*). Tokyo: Yadokari Shuppan.

Gorman, Rachel and LeFrançois, Brenda A. 2019. "Mad studies." In *Routledge International Handbook of Critical Mental Health*, edited by Bruce M.Z. Cohen, pp. 107–114. New York: Routledge.

Ingram, Richard. 2016. "Doing Mad Studies: Making (Non)Sense Together." https://journals.library.mun.ca/ojs/index.php/IJ/article/view/1680.

Ishihara, Kohji. (Editor). 2013. *Research of Tōjisha Research* (*Tōjisha Kenkyū no Kenkyū*). Tokyo: Igakushoin.

———. 2015. "Learning from Tojisha Kenkyu: Mental health 'patients' studying their difficulties with their peers." In *Disability Research Today: International Perspectives*, edited by Tom Shakespeare, pp. 27–42. New York: Routledge.

Ito, Takehiko. 2011. "Tōjisha kenkyū at Bethel House in Urakawa as a knowledge creation community: analysis via knowledge management theory (*Chishiki-sozo-kyodotai toshite no urakawa beteru no ie ni okeru tōjisha kenkyū—nareji managimento riron ni yoru bunseki*)." *Tozainanboku*, 110–129.

Kaneko, Juri. 2019. "Who is tōjisha?" *Glocal Tenri* Vol. 20 No. 1: 9 (one page only).

Kim, Dehuan. 2013. "The Current Situation of Tōjisha Kenkyu by People with Mental Disability in South Korea (*Kankoku ni okeru seishin shōgaisha ni yoru tōjisha kenkyū no genjō*)." *UTCP Uehiro Booklet*, No. 2: 175–182.

Kodaira, Tomoe, and Takehiko Ito. 2011. "Considering the ways of practicing '*tōjisha* as the protagonist': Using narratives by schizophrenia tōjisha as a clue (*'Tōjisha ga shujinkō' no jissen no arikata wo kangaeru: tōgōshicchoshō-tōjisha ni yoru naratibu wo tegakari ni*)." *Tozainanboku*, 130–140.

Kumagaya, Shin-ichiro. 2011. "Why Tōjisha? Why Research? (*Naze 'tōjisha' ka, naze 'kenkyū' ka*)." *Oral History Forum: Between Scientific Knowledge and Reality*, Special Issue 2: 93–100.

———. (2013). "An Invitation to Tojisha-Kenkyu: A New Science Focusing on Oneself" (https://www.u-tokyo.ac.jp/en/whyutokyo/indpt_tojisha_018.html).

Lewis, Bradley. 2009. "*Obsession: A History*; *The Loss of Sadness: How Psychiatry Transformed Normal Sorrow into Depressive Disorder*; *Shyness: How Normal Behavior Became a Sickness, and Bipolar Expeditions: Mania and Depression in American Culture* (review)" *Literature and Medicine*, 28 (1): 152-171.

Matsuda, Yasuko. 2018. "A critique of quality research on mental disability tōjisha experience (*Seishin shōgai tōjisha no keiken ni chakumokushita shituteki kenkyū ni kansuru hihanteki kentō*)." *The Bulletin of Research at Hokkaido University Graduate School, School of Education*. No. 132: 149–165.

McLelland, Mark. 2009. "The Role of the 'tōjisha' in Current Debate about Sexual Minority Rights in Japan." *Japanese Studies*, 29/2: 193–207.

McWade, Brigit, Damian Milton, and Peter Beresford. 2015. "Mad studies and neurodiversity: a dialogue." *Disability & Society*, Vol. 30, No. 2, 305–309.

Menzies, Robert, Brenda A. LeFrançois, and Geoffrey Reaume. 2013. "Introducing Mad Studies." In *Mad Matters*, edited by Brenda A. LeFrancois, Robert Menzies, and Geoffrey Reaume, pp. 1–22. Toronto: Canadian Scholars' Press.

Mukaiyachi, Ikuyoshi. 2006. *Anshinshite zetsubōdekiru jinsei (Peaceful acceptance of despair)*. Tokyo: Nihon Hōsō Shuppan Kyōkai.

———. 2009. *Gihō izen: Beteru no ie no tsukurikata (Before techniques: How to Build a Place like Bethel House)*. Tokyo: Igakushoin.

———. 2015. *Seishinshogai to kyōkai: kyōkai ga kyōkai de arutameni (Mental disability and the church: for the sake of being the church)*. Tokyo: Inochi-no-kotobasha.

Nakamura, Karen. 2013. *A Disability of the Soul: An Ethnography of Schizophrenia and Mental Illness in Contemporary Japan*. Ithaca, NY: Cornel University Press.

Nakanishi, Shoji, and Ueno, Chizuko. 2003/2015. *Tōjisha Sovereignty* (*Tōjisha Shuken*). Tokyo: Iwanami Shoten.

Ogura, Yasutsugu. 2014. "Tōjisha research and sociological power of infection" (*Tōjisha kenkyū to shakaigakuteki kannsenryoku*). *Mita Journal of Sociology* 19: 55–69.

Okuyama, Yoshiko. 2020. *Reframing Disability in Manga*. Honolulu, HI: University of Hawai'i Press.

Price, Margaret. 2011. *Mad as School: Rhetorics of Mental Disability and Academic Life*. Ann Arber, MI: University of Michigan Press.

———. 2015. "The Bodymind Problem and the Possibilities of Pain." *Hypatia* Vol. 30, No. 1: 268–284.

Sato, Mao. 2015. "To enable the true independence of disability tōjisha and their family (*Shōgai tōjisha to sono Kazoku — shin no 'jiritsu' wo kanaeru tameni*)." *Waseda University Culture Structure Studies Thesis*: 63–80.

Sato, Sonomi, Ikuo Mikaiyachi, and Kaori Okuda. 2017. "A research report on Open Dialogue (*Ōpun daiarōgu kenshū hōkoku*)" *The Bulletin of Health Sciences University of Hokkaido, School of Nursing and Social Services*, No. 24: 51–58.

Shinoda, Sayoko. 2001. "Who is the tōjisha in the case of domestic violence? (*Domesutickku vaiorensu ni okeru tōjisha towa dare ka*)" *Shakaiseisaku Kenkyu*, No. 2: 93–115.

Tateiwa, Shinya and Sugita, Shunsuke. 2016. *Sagamihara shōgaisha sasshō jiken: Yūseishisō to heitokuraimu* (The Sagamihara homicide of the disabled: eugenics and hate crime). Tokyo: Seidosha.

Teruyama, Junko. 2011. "*Hattatsu shōgaisha no katari kara: orutanatibu na 'tōjisha' sei ni mukete* (From the perspective of narratives on *hattasu shōgai*: toward an alternative tōjisha-ness)." *Tetsugaku*, 125: 313–338.

Toyota, Masahiro. 1998. "Tojisha-genso-ron. Aruiwa Minority no Undo niokeru kyodogenso no ronri (A Study of the Tojsha-phantasm: The logic of the shared phantasm in the minority movement)." *Gendai Shiso*, 26(2): 100–113. http://www.arsvi.com/1990/9802012.htm

Ueno, Chizuko. 2009. "Who is tōjisha?" *Kikan "Atto"* 15: 136–151.

Yoshii, Hiroaki. 2014. "On the potential of tōjisha 'research' in sociology: what is tōjisha 'research' aiming at? (*Tōjisha 'kenkyū' no shakaigaku no kaosei ni tsuite—tōjisha 'kenkyū' wa nani wo mezasu no ka*)." *The Journal of Mitaka Sociology*, 19: 70–79.

CHAPTER 3

Tōjisha Narratives

Tōjisha narratives—first-person narratives about lived experience of mental health distress—is the main focus of this chapter. As I explained in chapter "Tōjisha", tōjisha narrative (*tōjisha-gatari* or *tōjisha no katari*) is a genre of memoirs strongly associated with the tōjisha movement in Japan. The importance of the first-person perspective has been the key issue in social movements and identity politics. Equating self-narratives with tōjisha movements and identity theories, advocates of memoirs have argued that the subjective experience of tōjisha narratives on social issues possibly influences public opinion because other tōjisha in similar situations can easily relate to the narrators (e.g., Kido 2007).

In this chapter, the word tōjisha ("involved persons") continues to carry weight and be used to refer to individuals who identify themselves as psychiatric service consumers chiefly because the term has been adopted by them as a self-referent to rebuff the label of *seishin shōgaisha* ("mentally disabled persons") branded by medical professionals and the government. The chapter aims at answering the following questions: How did the genre of autobiography emerge and grow in the United States and Japan? What do tōjisha narratives accomplish in destigmatizing mental health conditions that other literary genres do not? Why study mental disability through the genre of autobiographical comics? A literature review of the genre of memoirs is presented in order to highlight its characteristics and

Y. Okuyama, *Tōjisha Manga*,
https://doi.org/10.1007/978-3-031-00840-5_3

strengths proposed by academics, including disability studies scholars. The reader will find a parallel in many characteristics between the English-language tōjisha narratives and their Japanese counterparts in this chapter.

First-Person Narratives

As tōjisha narratives are first-person narratives, this section provides the background of this literature genre. In Japan, we have the literary tradition of *shishō-setsu* (I-novels) as well as the term *jiden* (self-story), which refers to memoirs and other similar forms of autobiography.[1] In English, various labels, including autobiography, memoirs, life writing, life history, and autobiographical manifesto, are used for first-person narratives with slight differences in nuance and context. Other labels of first-person accounts, particularly in the context of academic research, are autoethnography (or personal ethnography), personal anthropology, self-research, self-ethnography, and complete member research.

Similar to the concept of *tōjisha kenkyū* described in the previous chapter, autoethnography is a form of personal narrative with an academic bent. Typically, the investigator is the central and only subject in autoethnographic research. Introduced in the 1980s, autoethnography entails this storytelling element of the researcher's self-narrative. Since then, mental health self-talks conducted in this academic approach have been increasingly accepted and valued at academic conferences and research publications in English-speaking countries (Foster et al. 2005). This design allows the researcher to play multiple roles; for instance, in examining experiences of psychiatric care, the investigator can be an academic investigator, a mental health professional, and a service user himself/herself (e.g., Liggins et al. 2012). Also called "insider research," autoethnography is, by definition, an ethnographical method in which the researcher is a member of the group under investigation and shares social as well as historical connections with that group; thus, the researcher undertakes both "insider" and "outsider" identities (Foster et al. 2005). Insider research is conducted from an emic perspective—an important aspect of anthropological research meaning the viewpoint from within the cultural group, as opposed to the etic perspective of the observer or outsider. Having a direct experience with the group being analyzed, the researcher is equipped with the emic perspective and can make a unique contribution to the research field.

Insider research has both shortcomings and strengths. For example, while being able to take advantage of one's close relationship with the group and shed light on issues that are normally considered taboo or access knowledge that only the insider is privileged to have, the researcher's overfamiliarization with the community may result in obfuscating the issue under investigation (Foster et al. 2005). However, in exploring mental health issues, the merits of the researcher's identity as someone with the same or similar condition to that of their research subjects outweigh a potential issue of lacking scientific objectivity. The investigator's intimate knowledge of the psychiatric condition possibly reduces the investigator-subject hierarchy and facilitates trust from the subjects which will encourage them to disclose their stories more comfortably. Together, both the researcher and subjects can reflect on how they turned their crises into life lessons and moved on to the future (Foster et al. 2005). Furthermore, when journal readers or conference audiences learn that others have survived a similar ordeal, they may find it liberating and have their isolation and loneliness alleviated (Foster et al. 2005).

Whether it is a research method or a literary genre, first-person accounts have received their fair share of criticism. One of the most negative commentaries may be that a first-person narrative is a "narcissistic" self-revelation that is undeserving of serious examination. In academia where an objective, expert voice is preferred over an intimate, personal voice, academic publications are biased toward research that manifests the voice of reason written from the third-person perspective where the author is an expert on the subject. However, I challenge that view by questioning if complete objectivity is attainable in any form of research. It is highly unlikely to be purely objective, as we interpret the data by drawing meanings based on our own particular life experiences. Furthermore, I argue that carefully selected autobiographical works can be found empowering by other tōjisha with similar emotional conditions. As demonstrated in the previous chapter, tōjisha's honest sharing of their personal experiences can change other people's lives for the better. Furthermore, as educational materials, tōjisha narratives can be more memorable to the reader than dry, academic papers written by third-party experts. In the two sections, we will examine the particular area of first-person narratives—life writing on disability and illness—and explore both the benefits from and drawbacks of tōjisha narratives in studying and teaching mental health issues.

Life Writing on Disability and Illness

Disability autobiography in which the author's disability or illness experience is depicted is a growing subfield of disability studies in the United States (Couser 2021; Reaume 2017; Rembis 2018). Terms similar to *first-person narratives* include *autopathography*, *disability memoirs*, and *life writing on disability and illness* to name a few. However, each term has a slightly different focus on the person's lived experience. For example, autopathography is a term used to refer to a first-person narrative about the experience of a serious illness such as cancer or depression. When labeled as disability memoirs, the author's condition is likely to be categorized as a physical, intellectual, or psychological disability. By contrast, life writing is a genre that bears on the representation of the narrator's disability or illness and provides authentic accounts of life history, separate from the other (imaginative) genres of narratives such as fiction and drama. Regardless of the label, these memoirs are "a nonfictional record or representation of actual human experience" (Couser 2021: 31) and play "a democratizing role in our culture" (41).

Thomas Couser, the founder of life writing in disability studies, began to notice narratives of illness and disability in the early 1990s, at which time no one appeared to be studying that particular type of memoir. During that era, the United States saw the HIV/AIDS epidemic, which was deemed a "death" sentence back then, and memoirs of AIDS victims were published by survivors or their relatives and drew in many readers. Since the 1990s, the genre has continued to proliferate in the United States (Couser 2021: 217). Interestingly, however, memoirs "did not get much respect as a literary genre" before that upsurge (Couser 2021: 26), yet it was this genre of autobiography that "played a powerful role in a minority rights movement" (19).[2]

Similarly, autobiographical accounts of the lived experience of psychological illness and treatment have been published in ever greater numbers, a phenomenon indicating the astonishing popularity of this particular genre (Hawkins 1999). The most published autopathographical works were about breast cancer, followed by the topics of prostate cancer and HIV/AIDS. Some works were about chronic illnesses and disabilities such as color blindness, and autobiographies about mental ill-health experiences, most notably bipolar disorder, were equally popular (Hawkins 1999).[3] From the mid-1980s through the 1990s, more and more patients were turning to autopathography books for insights into the same conditions they had and information on conventional or alternative treatments (Hawkins 1999; Rembis 2018). Compared to medical reports, which were impersonal records of symptoms written by medical professionals, pathographies were narratives rife with the authors' personal anecdotes; the authors were genuinely expressing their hopes, fears, and concerns. This phenomenon indicates that people felt depersonalized about medical treatment (Hawkins 1999).[4]

In this section, we will narrow in on the subfield of life writing by psychiatric system survivors in particular. One of the important objectives of such first-person narratives is to depathologize the conditions commonly lumped together and called "mental illness" (Wilson and Beresford 2002; Cahn 2014; Courser 2021). In the Western world, although the first-person accounts of psychiatric conditions such as the tale of Nebuchadnezzar in the Book of Daniel have been documented since biblical times, scholarly interest in first-person accounts of what mad studies scholars call "madness" has been growing since the mid-1980s (Reaume 2017; Rembis 2018). Two groundbreaking works were Dale Peterson's and Roy Porter's books in the 1980s that "helped develop a trend in which writings by mad people were published or republished in edited collections during the 1990s" (Reaume 2017: 281). According to Price (2011), the number of published memoirs about mental distress has proliferated in recent decades, and even the publishing industry acknowledges "the transgressive power of such life writing" in the United States (178). Writers of psychological disability memoirs make a valuable contribution to our understanding of mental health through an unconventional approach to "refiguring the rational" and add important perspectives to disability studies (Price 2011: 195).

Here, it is also important to point out cautions associated with first-person narratives on disability and illness raised by Thomas Couser, the leading voice of life writing as I mentioned earlier, as well as Geffrey Reaume, historian and mad studies scholar who specializes in first-person narratives of lived experience.[5] Due to the limited scope of this chapter on tōjisha narratives, I will highlight only some key points relevant to this book. First, because what we commonly call "mental illness" is a classification of the *Diagnostic and Statistical Manual of Mental Disorders* (hereafter *DSM*) that is "subject to social and cultural preconceptions" (Couser 2021: 205) and so-called diagnoses are only "lightly grounded in physiology" (209), we must be careful how to label the psychological condition of the narrator. Another caution he raises is that, as people with disabilities were historically exploited by the medical authority, writings labeled an "autobiography" are often associated with works written by ghostwriters including doctors and biographers (Couser 2004). The exploitation or misrepresentation of people with disabilities is one of the serious pitfalls inherent in life writing. As an example, in his article "Raising Adam" (1998), Couser critiqued Michael Dorris's *The Broken Cord*, especially how easily the parent's view is projected in a narrative accounting for the disabled child. Thus, Couser (2021) discourages the act of projecting onto the tōjisha any emotion that the author thinks they would feel in storytelling. This ethical code of conduct in life writing seems to be held strongly by the two family-caregiver authors of *tōjisha manga* in this book as they emphasized in their interviews that they always double-checked with their husbands, the tōjisha, while illustrating their lived experience of depression or OCD.

Another important point about disability memoirs is the lack of diversity influenced by gender, class, and education. As Reaume (2017) points out, prior to the twentieth century, first-person accounts, especially as primary sources, were nonexistent in "racially marginalized and culturally oppressed people throughout" the history of mad people's memoirs (284). To this day, inequality of access to literacy is real, especially in economically deprived regions of the world. This issue resonates when it comes to who has direct access to the art form in *tōjisha manga*: it is sometimes not the tōjisha but his spouse who is endowed with the privilege of drawing his direct experience professionally. (Thus, it is inevitable for the genre of *tōjisha manga* to include narratives by caregiver/family tōjisha.) Under those circumstances, it is critically important that the tōjisha and manga artist collaborate closely for the authentic representation of lived experience. The final point that seems pertinent to this book is how the societal perception of "madness" influences the credibility of the memoirist due to "the malleable nature of what madness meant at the time" in history (Reaume 2017: 287). Fortunately, all the five *tōjisha manga* featured in this book appear to be received with high credibility (e.g., film adaptation, awards) partially due to a growing awareness of the illustrated condition as described in each case study. As Reaume adds, first-person accounts through art are an increasing trend of life writing on mental health issues internationally. *Tōjisha manga* is certainly a good example of this trend.

Autobiographical accounts written by psychiatric patients, parents, and doctors-as-patients offer insider perspectives on the experience privileged only to them, telling what it is like to live through the stigma attached to the label of "insane." To illustrate this point, I will cite the mental health research conducted by Liggins et al. (2012) here. The first author of the study, Jill Liggins, is a psychiatrist who experienced depression for years and was admitted to The Hall, a mental health care unit that admits patients with conditions such as manic depression, schizophrenia, eating disorders, depression, anxiety, and alcoholism in New Zealand. The Hall is the nation's only acute care system after New Zealand's last major institution was closed about twenty years earlier. This autoethnographic paper was composed based on her diaries as a service user of mental health care. As the researchers analyze Jill's self-narrative, they demonstrate how autoethnographic writing can be both professional and personal, both analytical and reminiscent.

Although experiencing a serious illness is "a painful, disorienting, and isolating experience," by writing about their agonizing experiences, the authors are able to place themselves at the center of their narrative and view their illnesses and treatments in accordance with their perceived meaning of life (Hawkins 1999: 129). For many of them, writing an autopathography leads not only to discovering meaning but also strengthening their mental resilience. In other words, autopathography's role is to facilitate the writer's own recovery. In Hawkins's analysis, autopathography comes in four patterns: (1) didactic pathographies—narratives that are motivated by the explicit desire to help fellow humans and blend a personal account with practical information, (2) angry pathographies—narratives motivated by the desire to point out deficiencies in the areas of patient care, (3) alternative pathographies—narratives that are critical of the medical system but absent of angry denigration of medical professionals and provide alternative types of treatment, and (4) ecopathographies—a more recent type, narratives that link the author's experience of illness with broader political, cultural, or environmental issues (e.g., cancer is perceived as the product of a toxic environment) often with a "prophetic" tone of warning (Hawkins 1999: 128). This is an important and relevant insight as the majority of the *tōjisha manga* titles selected for this book are of the first type—didactic pathographies. That is, either in the introduction or the afterword of their manga, the authors of these graphic memoirs typically state their strong desire to help other tōjisha with similar conditions. I will now introduce Japan's literature genre of tōjisha narratives, *tōbyōki*, and show how "illness," including mental distress, has been depicted in that tradition.

Tōbyōki (*Illness Memoirs*)

Recently, there has been an increase in the publication of tōjisha narratives about various social issues in Japan. Previous research has demonstrated that tōjisha narratives actually help reduce stigma. For example, Sakae (2016) argues that sharing self-narratives through public talks is a powerful way for tōjisha to make social changes. Kido (2007) equates self-narratives with tōjisha movements and identity theories and argues that the subjective experience of tōjisha narratives on youth labor issues has a strong potential to influence public opinion because other youth in similar situations can easily relate to the narrators. The importance of the I-perspective has been the key issue in social movements and identity politics.

The most well-documented field of research on tōjisha narratives is the study of *tōbyōki* (illness memoirs, illness narratives), a literary genre that has grown rapidly since the late 1990s in Japan (Hongo 2013; Kadobayashi 2011), which coincides with the time when doctors began to notify patients of cancer diagnosis more commonly (Noguchi 2018). Literally meaning "battling illness," the word *tōbyō* was coined by a tuberculosis patient, Kosakai Fuboku, for the first time in the 1920s (Kadobayashi 2011; Noguchi 2018). Kosakai was a medical doctor who contracted the illness when he was studying abroad and promoted the spirit of "fighting the illness" as the proper way of strengthening one's will to live while undergoing medical treatment (Noguchi 2018: 103). Upon returning to Japan, he publicized his ideology of *tōbyō* through two books *Tōbyō-jutsu* (*The Art of Fighting the Illness*) in 1923 and *Tōbyō-mondō* (*The Q&A of Fighting the Illness*) in 1924. Although a similar label, *tōbyō-shuki* (illness diary) exists, *tōbyōki* will be used as an umbrella term for pathography, or patients' narratives of illness, in this section.

Similar to the US genre of patients' narratives of illness discussed in the previous section, *tōbyōki* are accounts written by tōjisha (or sometimes by their families). However, unlike their US counterpart, Japan's autopathography is deemed a useful resource for medical education, which is, in fact, many *tōbyōki* authors' primary reason for publication (Kodaira and Ito 2009). The publication of *tōbyōki*, especially memoirs of mental distress, has surged recently, as exemplified by the many autobiographical narratives of illnesses in various forms of popular culture, including manga (Kodaira and Ito 2009). The genre's popularity in Japan is due to the fact that tōjisha narratives offer valuable information, especially to the supporters of tōjisha, including their family and friends. Books of the *tōbyōki* genre are displayed distinctively in the front area of bookstores and are frequently covered by the mass media; in the fields of nursing and psychology, disability memoirs have frequently been incorporated into qualitative research and *tōbyōki* books published under the author's real name are featured in presentations at nursing studies conferences; the libraries categorize *tōbyōki* as a special category in their Internet search system (Kodaira and Ito 2008, 2009, 2011). As the field of mental health welfare in Japan has made historical changes, tōjisha narratives began to feature progressive approaches such as early intervention (*sōki shien*), support for employment (*shūrō*), and self-transformation (*jiko hen-yō*) (Matsuda 2018). Currently, the newer generations of psychiatrists and psychologists are being trained on the ICD-11 manual published by the World Health Organization (WHO) in Japan. According to its guidelines, the diagnosis of a mental disability must derive from examining the patient's whole life, validating the use of disability memoirs and validating their positive effects on mental health care and patients' healing processes. It is also important to empathize that the genre of *tōbyōki* does not distinguish the authors who are family caretakers (*kazoku tōjisha*) from those with lived experience (*honnin tōjisha*) based on the degree of tōjisha-ness (see Kodaira and Ito 2011 and Yagi 2009, for example). This fact further supports my position of clustering the two closely related groups in the tōjisha category.

As with one of the purposes of writing life articulated by Couser (2021) in the previous section, the literary genre of *tōbyōki* is a "public narrative" also designed to reduce social stigma against people with mental disability (Matsuda 2018). This merit of tōjisha narratives—the reduction of stigma—strongly resonates with *tōjisha manga*. Research shows that stigma, especially self-stigma, is harmful to individuals with serious psychiatric conditions such as schizophrenia because they are more likely to internalize stigma compared to those with other disabilities and thus experience diminished self-esteem (e.g., Watson et al. 2007). Therefore, the reduction of self-stigma helps tōjisha achieve their psychological well-being.[6] Besides the reduction of stigma, other merits of tōjisha narratives include the promotion of community-level mental health awareness and the tōjisha's decreased risk of relapse (Matsuda 2018). In connection with mental health issues, which is the focus of this book, several studies below shed light on what makes tōjisha narratives so special.

For instance, Yagi pioneered the promotion of using *tōbyōki* as a training resource for nursing students more than ten years ago. A psychiatrist specializing in schizophrenia, Yagi (2009) calls for restoring tōjisha's rights and reputation in his book *Learning About Schizophrenia from First-Person Narratives* through utilizing autobiographies by people with lived experiences of schizophrenia. That is because, in his view, only the tōjisha—whether the person with the condition or their caregiving family—has the privilege to write about their experiences. Organizing approximately twenty *tōbyōki* books by themes, he reviews each narrative's depiction of the inner world of schizophrenia. Although the reader benefits from accessing an invaluable insider perspective, the writer gains much from the experience of writing a *tōbyōki*. As one of the common characteristics of this genre, the writers state that their lives are somewhat "improved" by the process of self-reflection. Even though it means that their private trial of hardship is publicly disclosed, Yagi argues, the tōjisha feel accepted as equal human beings and have their existence validated through audience feedback and reception. Personal gains and acceptance are important traits of tōjisha narratives as the featured manga artists also expressed very similar sentiments in my interviews.

However, it is not easy to illustrate one's lived experience of emotional distress in *tōbyōki*, especially because it is hard to find exact words that accurately describe the experience that may be so frustrating and confusing at first. Interestingly, Sawada (2014) describes the process of finding what to say and how to explain things through one's repeated efforts of storytelling as seen in the examples of *tōjisha kenkyū* discussed in the previous chapter. The tōjisha become better at illustrating their conditions with words by using metaphors creatively, Sawada explains. For example, although it is hard to communicate your emotional pain because it is so subjective, when asked by a non-tōjisha person, "How does that pain feel to you?," you may come up with analogies like "heavy blankets" or "as if being pulled by a wire-like, firm thread." Such creative narration with common images that anyone can understand, he argues, helps make the subjective experience of psychological pain more comprehensible to others. Especially in regard to one's inner state, ordinary words like "sad" cannot fully describe in what manner the narrator is affected by the feeling. Thus, one strategy to compensate for the lack of precise words is to use metaphors; another strategy is to borrow each other's creative phrases used in peer communication. As I will show the reader in the latter part of this book, the use of metaphors and other nonstandard expressions in tōjisha narratives strongly resonates with unique visual strategies employed by the *tōjisha manga* artists when representing their psychological distress (e.g., a depression onset is metaphorically drawn with a creepy-crawly slithering slowly approaching the character). Similar to *tōbyōki* authors, the *tōjisha manga* artists also appear to "borrow" visual expressions from each other. Another common technique appropriated by the *tōjisha manga* artists is personification (*gijinka*), the tactic of personifying an abstract thing. These creative solutions for "the lack of precise words" will be presented in the analysis chapters.

Memoirs of disability and illness typically tell the reader about a journey through which the tōjisha come to make sense out of their lived experience or about particular encounters with peers that have served as the turning point in their lives. In such memoirs, they narrate the process of accepting the condition and reinventing their personal history. Similarly, *tōbyōki* tells the reader about the journey of acceptance and breakthrough moments of their lives (Matsuda 2018). While the proliferation of tōjisha narratives in the genre of *tōbyōki* helps diversify tōjisha's demographics as well as research methods in the academic realm, it also yields the political empowerment of the tōjisha (Matsuda 2018). Furthermore, as the field of mental health welfare has made historical changes, new themes such as early intervention (*sōki shien*), support for employment (*shūrō*), self-help group, peer support, self-transformation (*jiko hen-yō*), and recovery began to appear in tōjisha narratives. Another merit of publishing tōjisha narratives is reduction of self-stigma, which I believe also speaks for *tōjisha manga*. Research shows that self-stigma is harmful to tōjisha, especially individuals with serious conditions such as schizophrenia, because they tend to internalize stigma more than those with other disabilities and thus experience diminished self-esteem (e.g., Watson et al. 2007). Thus, reduced self-stigma contributes to the development of tōjisha's psychological well-being. Other merits of tōjisha narratives are reduction of societal stigma, promotion of community-level mental health education, and decreased risk of relapse (Matsuda 2018).

Much like Matsuda's point, Sakae (2016) also argues that *yamai no katari* (the talk of the illness) is a great tool for empowerment. Her paper reports on what she found from semi-structured interviews with two progressive members of Piano, a mental health volunteer group that organizes tōjisha's public talks to empower other individuals with lived experience and promote social change. The group helps arrange two types of tōjisha activities: self-narrations open to the public and peer-group self-narrations. To her, disease is a biomedical or psychosocial process problem while illness is the person's experience of the perceived disability and meaning brought by the experience. This frame of mind allows us to reconceptualize the meaning of lived experience, not in a medical way but rather in a personal way. Thus, she argues, to reframe and make sense of our lived experience, we must engage in the storytelling of *yamai no katari* in which the tōjisha is the protagonist telling the tale from their point of view. Those who write *yamai no katari* help bring social awareness about oppression and prejudice against people with lived experience, empower those who share the same sentiments, and encourage other tōjisha to engage in activism. In other words, sharing the tale with others and influencing their thoughts and emotions enables the tōjisha to transform themselves, too.

From the interviews, Sakae (2016) learns that an honest story (*arinomama no monogatari*) is better suited for peer-group presentations while a recovery story with a happy ending (*rikabarii no monogatari*) is better received in public talks whose audiences include both tōjisha and non-tōjisha. When sharing their own painful illness experiences with peers, it is easier to establish trust and comfort. Thus, some Piano members prefer peer-group talks. Public talks, by contrast, require extra creative efforts to sustain the audience's interest, especially with school-age listeners. Depending on the type of audience, presenters also need different strategies, such as the use of storyboard tools, to discuss medical information.

One of Satake's interviewees, Mr. Mori, shares his *yamai no katari* to "enlighten," in his words, the public about this most misunderstood illness.[7] Seventy percent of his public talk is about his experience of schizophrenia, and the other 30% consists of his messages to the audience. To this day, recalling the awful events he experienced makes him cry; thus, he does not share the most painful parts of his experience. He believes that he should share whatever feels comfortable and should not be forced to disclose everything. Mr. Mori utilizes multiple strategies to express "invisible symptoms of schizophrenia" (*kashika dekinai tōgōshicchō no shōjō*), one of which is using manga-like illustration to show what auditory hallucinations (*genchō*) and delusions (*mōsō*) are like. He said in the interview as follows:

> The discourse of illness tends to be dark and depressing. The audience gets depressed if all they hear is a sad story. So, my motto is to keep it positive, entertaining, and humorous. For that, I incorporate manga, sing with the audience for twenty minutes or so when their attention wanes, and conclude my talk with Ed Harumi's *Gu-gu* dance. That's what entertainment is all about, right?

Although the general public's understanding of depression has been improving, schizophrenia is still poorly understood in Japan. Mr. Mori was well versed in public speaking before being diagnosed with schizophrenia and uses that experience to increase public awareness about this sparsely understood psychiatric condition. Besides public talks, he enjoys writing essays. He states that, although he writes with his auditory hallucinations in tandem (*genchō to ninin-sankyaku de genkō wo kaku*), he is satisfied with getting catharsis through writing (*kaku kotode katarushisu ga eraretakoto mo ōkii*). Writing has a detoxing effect, freeing him from the stress of muddy and painful experiences. It also enables him to see his situation objectively, which helps him gain a metacognition (*meta-ninshiki*) of his lived experience. This metacognition comes to him, especially when he tries different ways to make his condition comprehensible to the general audience. For example, he uses the *katakana* (phonetic script for loan words and onomatopoeia) to separate his non-delusionary thoughts from the voice of delusion he hears concurrently. Such an unconventional way helps the reader imagine his experience of auditory hallucinations.

To educate people about this least understood illness, Mr. Mori circumvents the risk of exposing his painful experience to unspecified audiences by inventing unique strategies and making the most of his sui generis entertainment talent. He believes taking this risk is worth it if he can contribute to breaking down the barrier between the oppressive society and people with lived experience like him. Of course, such a choice may not be affordable for every tōjisha. However, becoming a tōjisha narrative pioneer of his generation is what he chose to do. He continues to invent new narrating methods and polish his storytelling skills. Mr. Mori has become a public speaker in high demand and is now considered an "entertainer" who has been interviewed as a guest along with people who are well recognized in their fields. Individuals diagnosed with schizophrenia are not "weak" patients or "disabled" persons that society needs to defend. Rather, he believes, they are people who fight and capitalize on their auditory hallucinations and delusions to press the oppressive society to change. To this end, he encourages his schizophrenic peers to share their stories in public. It is amazing how much Mr. Mori's choice of an "audience-friendly" disability memoir and use of humor resembles the way manga artist, Azuki Daruma, characterizes her autobiographical manga as reported in Nishikura's (2017) "Humor for Survival," which will be presented in the next chapter on autobiographic comics in Japan.

Like Yagi (2009) and Matsuda (2018), Kodaira and Ito (2008, 2009, 2011) have examined a wide variety of *tōbyōki* publications to summarize the genre's benefits. The narratives of *tōbyōki* are first-person accounts that inform the reader how the authors confronted and eventually came to terms with their own mental health issues in intimate, and sometimes self-deprecating, manners (Kodaira and Ito 2008). Two good examples of *tōbyōki* these researchers recommend in this 2008 paper are the trilogy about schizophrenia by Furukawa (2001, 2004, 2006) and a doctor-as-patient memoir by the psychiatrist Arizuka Ryoji (2005) about his depression.[8] In my view, these two representative *tōbyōki* stories have many things in common with the selected *tōjisha manga* in terms of narrative characteristics. What separates Furukawa's and Arizuka's books from medical books on psychological conditions is the strong presence of the tōjisha point of view in the narrative. Furthermore, the author's humorous writing style not only invigorates the reader but also provides "*datsuryoku-kei anshin*" (relief with slice-of-life comedy).

Previous research has shown that to reduce the viewers' prejudice against mental health issues and narrow social distance between the tōjisha and non-tōjisha, *tōbyōki* documentaries are more effective than educational videos featuring medical doctors on the same subject, especially when used in the classroom. Because tōjisha narratives offer benefits that are not necessarily obtained through educational documents with psychological statistics and anonymous patients' case studies, *tōbyōki* are far more persuasive and useful as classroom material and should be considered a social and medical resource and a tool for empowerment (Kodaira and Ito 2008). From the tōjisha movement point of view, publishing tōjisha narratives is a way to amend the wrongs done to this minority group who formerly were locked up in hospitals and excluded from society. In this regard, *tōbyōki* entails the power of social change, turning Japan into a fertile ground for cultural diversity. This genre has filled an important role in contributing to society by providing the public with an opportunity to properly understand the nature of *kokoro no byōki* ("a sickness of the mind," a common Japanese euphemism for a mental disability); therefore, it should be promoted as teaching materials (Kodaira and Ito 2009).

As the other researchers cited in this section point out, Kodaira and Ito (2009) also highlight the symbiosis between the writer and reader in disability memoirs: The writer benefits from reflecting upon their experiences and feels restored, and the reader learns the insider perspective of how the illness impacts a person's life. They state that, like attending *tōjisha kenkyū* workshops, non-tōjisha can vicariously experience the narrator's emotions in reaction to their uncontrollable condition by reading *tōbyōki*. The benefits that the tōjisha receive are threefold: a sense of belonging (*shozokukan*), self-acceptance (*jiko juyō*), and a degree of contribution (*kōken-do*). For instance, the kind of "shared community" formed between the writer and reader would serve as an *ibasho* (place of belonging) for the tōjisha. Through the *ibasho*, the tōjisha feel accepted by others. This community building is crucial for the tōjisha to feel accepted and included in the society and eventually be ready to accept who they are. Furthermore, through this established community, information as to where people with mental health issues can receive adequate support becomes more accessible to them. In this regard, the formation of such a community provides the tōjisha with the aforementioned three benefits: belonging, acceptance, and social contribution. Now that they know they are producing something valuable, the tōjisha also become aware that their life does not end after being publicly branded with the contested label of a *seishin shōgaisha* (mentally disabled person). The same outcomes can be expected of people with mental health problems who are enrolled in the *deikea sentā* ("daycare center"), facilities for seniors or disabled persons that offer short-term care services during daytime hours in Japan (Kodaira and Ito 2011). The *deikea sentā* becomes the tōjisha's *ibasho*, a place of belonging, relieving them from total isolation.[9] Regardless of the specific location of connection with the external world, once their *ibasho* has been established, people with schizophrenia can begin their own process of searching for how to live.

As the merit of this genre, *tōbyōki* forces narrators to confront and process the meaning of their illness and the strategies they chose to deal with it, which inherently has the effect of healing for themselves. However, it should be emphasized that, although *tōbyōki* can be healing, searching for the right words to describe one's lived experience can take an enormous amount of energy from the narrators, especially if they are still in the process of recovery (Kodaira and Ito 2009, 2011). That is why most tōjisha narratives are written after the author has reached a certain level of mental stability (Kodaira and Ito 2009). The issue of the prime time for writing is a topic relevant to my case studies of the *tōjisha manga* titles selected for this book. It is also important to mention that writing a tōjisha narrative, even after the time of trauma, is not totally stress-free as Hosokawa Tenten, author of *My SO Has Got Depression* (*Tsure ga utsu ni narimashite*), noted in the first volume; she recalled how dealing with her husband's depression was much harder than she had thought it would be.

Interestingly, in their 2008 paper analyzing several examples of *tōbyōki*, Kodaira and Ito take a close look at "comics of mental health memoirs" in Japan, the exact subgenre of autobiographical comics this book intends to delineate. Similarly, in their 2011 paper, Kodaira and Ito analyze *tōbyōki* comics authored by either people with lived experience of schizophrenia or their family caretakers. Regardless of the format, an important feature of the *tōbyōki* is that the narrator is also the protagonist of the story, a characteristic also shared with *tōjisha kenkyū* and most importantly, *tōjisha manga*. To segue into the next chapter on Japan's graphic memoirs, this chapter will end with a discussion of autobiographical comics on illness and disability that are increasingly gaining attention from scholars outside of Japan.

Representations of "Illness" in Autographical Comics

Comics have been popular since the nineteenth century, and both North America and Japan have "mature comic markets" (Chu and Coffey 2015: 145). Art Spiegelman's (1986) *Maus* and Alison Bechdel's (2007) *Fun Home* are some of the most frequently identified masterpieces that examine stigma/trauma graphic memoirs published in Western countries (e.g., Chute 2017; Kunka 2018). In North America and England, the genre of autobiographical comics has garnered scholarly attention, as seen in publications such as Hillary Chute's (2017) *Why Comics?*, Andrew J. Kunka's (2018) *Autobiographical Comics*, and Frederik Byrn Køhlert's (2019) *Serial Selves*. Scholars in comic studies such as Chute (2017) assert that comics is "a medium in its own right—not a lowbrow genre of either art or literature" and is "an art form with an ever-increasing popularity and range" nowadays (2). Autobiographical comics are placed in the caliber of comics for adult readers by these researchers. Yet, as Chute cautions, the genre is still mistakenly associated with comics about superheroes by non-experts. This section presents a literature review of autobiographical comics to validate the legitimacy and preeminence of this genre and argues for its cultural significance. Readers interested in a historical account of autobiographical comics and a more detailed discussion of the genre's characteristics in the West should turn to those books by Chute (2017), Kunka (2018), and Køhlert (2019).[10]

The US prototype of autobiographical comics emerged as underground comics. Chute (2017) characterizes the underground comics revolution that began in the 1960s and 1970s mainly in New York and San Francisco as "a vital movement with literary, artistic, and popular force that profoundly shaped contemporary comics" (13). Unlike other comics, underground comics were produced outside of regular commercial strictures and were often labeled "comix" with an *x* for an adult edge. The first comix work was R. Crumb's *Zap #1*, and after the success of the first issue, the cartoonist opened the venue to other underground comic artists. This underground comic movement was associated with the counterculture of the United States in the late 1960s. The underground comic artists produced complex, nuanced, and sometimes absurdist storylines. Because the target readership was adult intellectuals, underground comics were different from the regular, superhero comics; as Chute (2017) explains, they were "comics without punchlines" and were "created with complete artistic and intellectual freedom to represent an artist's personal vision" for adults (13). It was in this genre of comics that *Binky Brown Meets the Holy Virgin Mary,* Justin Green's influential work, was published in 1972. This work is considered the first autobiographical comic by comic studies scholars as well as comics creators (Chute 2017; Kunka 2018).

Chute (2017) states that graphic memoirs of illnesses and disabilities "use the show-and-tell aspect of comics basic to its hybrid form to reveal hard-to-convey truths about sickness or ability" and "have multiplied hugely over the past twenty years" (240). When it comes to mental disability, my Internet search yielded several recently published autobiographical comics, including Ellen Forney's *Marbles: Mania, Depression, Michelangelo, and Me* (bipolar disorder) in 2012; Allie Brosh's webcomics, eventually published as a comic book *Hyperbole and a Half* (depression) in 2013; Darryl Cunningham's *Psychiatric Tales: Eleven Graphic Stories About Mental Illness* (healthcare assistant) in 2013; Katie Green's *Lighter Than My Shadow* (anorexia nervosa) in 2017; Rachel Lindsay's *RX* (her experience with the advertising and treatment of psychiatric medicines) in 2018; and Chuck Mullin's *Bird Brain* (about his anxiety and depression) in 2019.

Both Chute (2017) and Myers and Goldenberg (2018) shed light on a growing movement, "graphic medicine." This movement seeks to use *graphic pathographies*—comics that accurately describe the experience of illness—as powerful resources for medical education and patient care. According to Myers and Goldenberg, this movement offers "important implications for the emphatic practice of whole-person medicine" (165). Medical practitioners and trainees find in these comics new insights into the lived experience of mental ill-health and understand it in its complexity. From the comics, people with lived experience gain new information, feel less isolated, and learn to cultivate their own "practical skills that might enhance their autonomy and moral agency" (165). The movement echoes "collaborative autobiography," which Sklar (2012) promotes. This mode of autobiography uses comics as a means for the intellectually disabled to talk about their disability issues. Although it is beyond this book's capacity to discuss further, it is important to mention that there are many more approaches to utilizing self-portrayal comics in the West. With regard to autobiographical comics on illness and disability, Chute (2017) writes that comics can "richly present the consciousness of a person suffering from a mental disorder" (240). In addition, these comics are capable of making visible both external (e.g., symptoms) and internal features (e.g., cognitive and emotional impacts) that are otherwise hard to convey accurately. In the United States, the first autobiographical comic on illness and disability is Justin Green's underground comic, *Binky Brown Meets the Holy Virgin Mary*, published in 1972. The story tells of his own experience with debilitating OCD. Green's pioneering book has set the stage for more graphic autobiographies of this sort to come. For example, Stack et al. published *Our Cancer Year* in 1994; David B.'s (2005) *Epileptic*, a graphic memoir about his brother's epilepsy, came out in 2005. More recently, Cece Bell's deaf comic, *El Deafo*, and Peter Dunlap-Shohl's *My Degeneration: A Journey Through Parkinson's* were published in 2014 and 2015, respectively.

Interestingly, Myers and Goldenberg (2018) found readers better recall the information from graphic pathographies than from medical literature. They assert that graphic pathographies are an excellent means to understand from the patient's point of view what it is like to be afflicted with a disability. Comics have other advantages as a medium such as appearing nonthreatening and inviting, particularly because of comics' association with pleasure reading. For patients and family members, this medium is a great match. Furthermore, medical students have severe time constraints with heavy reading assignments and duties on the wards, to say the least; because comics' words and images complement each other unlike dense text with many words and complex sentences in medical literature, they may find comics user-friendly with clear, quick, and easy-to-understand ideas. Despite being considered a medium of popular culture, Chute (2017) argues that these autobiographical comics on illnesses and disabilities excel at depicting the complicated experience of illness quite accurately and vividly. In the Japanese context, the genre of autobiographical comics is called *essay manga*, from which the subgenre, *tōjisha manga*, has derived. In the next chapter, I will describe essay manga as a Japanese comic genre to provide a non-Western examination of autobiographical comics to supplement the English-language books that discuss the genre almost exclusively from an Anglophone point of view.

Conclusion

This chapter first discussed the literature genre of autobiography. It then zoomed in on first-person narratives of disability and illness, a genre that has grown rapidly since the 1990s in both the United States and Japan. That section first touched upon a narrative genre of life writing on disability and illness in the United States and proceeded to introduce Japan's *tōbyōki* (illness narratives), a literature genre that began in the 1970s and proliferated in the late 1990s through the 2000s in Japan.[11] The chapter ended with a discussion of autobiographical comics, especially graphic memoirs of illness as a unique and noteworthy genre to segue into the next chapter on essay manga—Japan's equivalent of autobiographical comics, which is the genre to which this book's *tōjisha manga* belongs.

Notes

1. The genre *jiden* shares such characteristics as humor, sometimes gallows humor, the presence of supportive people, and the happy ending that are also some of the common characteristics of the selected *tōjisha manga* titles. An example of *jiden* is Shigematsu Kiyoshi's (2002) *Kiyoshiko*, a memoir about his childhood as a stutterer, in which Shigematsu recalls his painful interactions with schoolmates and contrasts that experience with his imagined, stammer-free chats with his imaginary friend "Kiyoshiko" whose name was taken from the Christmas song, *Kiyoshi konoyoru* ("Silent Night").
2. Reaume (2017) also adds that in Canada and the United States, academics began to publish on "the history and current status of the mad movement" instead of medical history perspectives in the early 2000s.

3. According to Hawkins (1999), the authors of autopathographies in the 1990s are more likely to be middle-class, Christian, public figures (journalists, athletes, ministers, etc.), including movie stars. Such "celebrity memoirs" are more likely narratives with the "overcoming" theme based on the medical model of mental disability (e.g., Brooke Shields's 2006 book, *Down Came the Rain: My Journey Through Postpartum Depression*) and have increased in publication in the past thirty years (Reaume 2017). However, in some cases, mental health memoirs by non-public figures have also become bestsellers. One such example cited by Couser (2021) is Susanna Kaysen's memoir, *Girl, Interrupted* (1993), based on the author's experience in a psychiatric hospital confinement after being diagnosed with borderline personality disorder. Although she is still considered "the progenitor of the now common mental-illness memoir," the author never expected her narrative would become a mega hit and be adapted into a popular movie, a seemingly surprising yet somewhat expected outcome propagated by the Clinton administration's antidiscrimination laws against psychiatric service users and the publication of *Listening to Prozac* by Dr. Perter Kramer as a criticism of the psychiatric system (Merrigan 2018).
4. Furthermore, Hawkins (1999) believes that autopathographies are also helpful for physicians: Doctors find little or no time to get to know their clients these days but are still expected to be aware of the patients' wishes, needs, and fears, and autopathography books provide them with an opportunity to obtain the patient's point of view and a longitudinal observation of the illness (because the memoirs tend to include their conditions both before and after the treatment). Similarly, he argues, patient narratives help physicians understand common complaints that patients have (e.g., how physicians fail to express empathy for them; dilemmas from receiving conflicting pieces of advice from different medical experts). Moreover, patient perspectives also inform physicians of alternative methods and the cultural attitudes or assumptions about the illness, treatment, or recovery that patients have, encouraging them to explore unconventional ways to help their patients fight the illness. Seen in this light, autopathographies are, in Hawkins's (1999) words, a veritable gold mine of patient attitudes and assumptions about all aspects of their illness. This view seems applicable even in the millennium as Couser (2021) states that narrative medicine is gaining strong support from medical practitioners. It is only natural that autopathographies continue to be in high demand, many of them bestsellers.
5. Upon reviewing my first draft of this chapter, Michael Rembis told me that he has been working on a monograph on the historical documentation of autobiographies by mad people. However, as he informed me, although it would be "super helpful" in my thinking on this book (email on September 30, 2021), his book, *Writing Mad Lives—In the Age of the Asylum*, will not be in print until 2023 (email on July 12, 2022). Thus, it is unfortunate that

I am unable to refer to his scholarship on mad people's memoirs in this chapter.

6. From a disability studies point of view, it is also critically important to increase public awareness of mental illness stigma, to reduce the tōjisha's internalized self-stigma. As shown in this chapter, Kodaira and Ito (2011) emphasize the importance of disseminating tōjisha narratives to non-tōjisha audiences.
7. In *Reframing Disability in Manga* (2020), I pointed out that some of the media's common portrayals of the disabled are "good cripples," "superscripts," as well as "victims" and "villains," as criticized by many disability studies scholars. We need to be careful not to see Mori's public speech as a self-portrayal of a plucky, courageous "overcomer," especially because he shares with the public not only his achievements but also his current struggles.
8. Arizuka's (2005) *tōbyōki* is titled *Utsubyō wo taikenshita seishinkai no shohōsen: Ishi toshite, kanja toshite, shiensha toshite* (*Prescriptions by a psychiatrist who experienced depression: as a medical doctor, as a patient, as a supporter*). As a psychiatrist who developed major depression and moved to a peaceful rural place in Okinawa for his new life, Arizuka offers treatment suggestions as well as prevention measures from the tōjisha point of view. While providing professional medical information about depression, he illustrates his lived experience of a mental health crisis in ways that make it comprehensible enough for those with no experience and offers his own unorthodox strategies for recovery. He presents his opinions about medical biases and highlights the shared goals of many people diagnosed with depression: how to accept and live with this condition as well as how to endure unwanted reactions from his surroundings.
9. Such positive effects of the *deikea* are depicted in the *tōjisha manga*, *Wagaya no haha wa byōki desu* (*My Mother is a Sicko*), by Nakamura Yuki.
10. In this chapter, my primary focus in the literature review is on the works of comic studies scholars both in the West and Japan as they are well versed in the semiotic analysis of autobiographical comics. For the literature review pertaining to critiques on disability representation in popular culture by disability scholars, see Chapter 2 of *Reframing Disability in Manga* (2020). *Disability in Comic Books and Graphic Narratives* (2016) is also recommended as a book that explores the representation of disability in comics from a disability studies perspective.
11. When I presented a talk about *tōjisha manga* at the early stage of this manuscript, one of the comments I received and found helpful was that I should examine the root of memoirs beyond the field of manga. Also, when Michael Rembis read the early version of this chapter, he recommended that I look into life writing from disability studies perspectives. Thus, this chapter includes my scholarly responses to the feedback by both.

References

Arizuka, Ryoji. 2005. *Utsubyō wo taikenshita seishinkai no shohōsen* (*Recommendations from a Psychiatrist Who Experienced Depression.*) Ōtsuki Shoten: Tokyo, Japan.

David, B. 2005. *Epileptic.* London: Jonathan Cape.

Bechdel, Alison. 2007. *Fun Home: A Family Tragicomic.* Mariner Books: Boston, MA.

Bell, Cece. 2014. *El Deafo.* New York: Amulet Books.

Brosh, Allie. 2013. *Hyperbole and a Half: unfortunate situations, flowed coping mechanisms, mayhem, and other things that happened.* New York: Gallery Books.

Cahn, Susan. 2014. "Border Disorders: Mental Illness, Feminist Metaphor, and the Disordered Female Psyche in the Twentieth-Century United States." In *Disability Histories*, edited by Susan Burch and Michael Rembis, pp. 258–282. Champaign, IL: University of Illinois Press.

Chu, Kin-wai, and Simon Coffey. 2015. "Multimodal Analysis of Graphic Novels: A Case Study Featuring Two Asian Women Travelers," *Intercultural Communication Studies* XXIV (1): 145–165.

Chute, Hillary. 2017. *Why Comics?: From Underground to Everywhere.* HarperCollins: New York, NY.

Couser, G. Thomas. 1998. "Raising Adam: Ethnicity, Disability, and the Ethics of Life Writing in Michael Dorris's 'The Broken Cord'." *Biography*, 21(4) (Fall): 421–444.

———. 2004. *Vulnerable Subjects: Ethics and Life Writing.* Ithaca, NY: Cornel University Press.

———. 2021. *The Work of Life Writing: Essays and Lectures.* New York: Routledge.

Cunningham, Darryl. 2013. *Psychiatric Tales: Eleven Graphic Stories About Mental Illness.* London: Blank Slate.

Dunlap-Shohl, Peter. 2015. *My Degeneration: A Journey Through Parkinson's.* University Park, PA: The Pennsylvania State University Press.

Furukawa, Natsuko. 2001. *Kokoro wo yamutte dōiukoto?* (*What does it mean to have the mind's illness?*) Budosha: Tokyo, Japan.

———. 2004. *Kokoro ga yamu toki, kokoro ga naerutoki.* (The time of the mind aching, the time of the mind withering). Budosha: Tokyo, Japan.

———. 2006. *Kokoro wo yamu hito to ikiru kazoku.* (*The Family of a Mentally Ill Person*). Budosha: Tokyo, Japan.

Foster, Kim, Margaret McAllister, and Louise O'Brien. 2005. "Coming to Autoethnography: A Mental Health Nurse's Experience." *International Journal of Qualitative Methods*, 4(4): 1–15.

Forney, Ellen. 2012. *Marbles: Mania, Depression, Michelangelo, and Me.* New York: Avery.

Green, Justin. 1972. *Binky Brown Meets the Holy Virgin Mary. Last Gasp Eco Funnies*: San Francisco: CA.

Green, Katie. 2017. *Lighter Than My Shadow*. St. Louis, MO: ROAR.

Hawkins, Anne Hunsaker. 1999. "Pathography: patient narratives of illness." *Culture and Medicine*, Vol. 171 August: 127–129.

Hongo, Masatake. 2013. "Review of Kadobayashi Michiko's *For Empowerment—The Sociology of Cancer Memoirs* (*Kadobayashi Michiko cho, Ikiru chikara no minamoto ni—Gan tōbyōki no shakaigaku*)." *The Japanese Society of Health and Medical Sociology*, 24(1): 74–75.

Kadobayashi, Michiko. 2011. *For Empowerment—The Sociology of Cancer Memoirs* (*Kadobayashi Michiko cho, Ikiru chikara no minamoto ni—Gan tōbyōki no shakaigaku*). Seikaisha: Japan.

Kosakai, Fuboku. 1923. *Tōbyō-jutsu* (*The Art of Fighting the Illness*). Shunyodo: Tokyo, Japan.

———. 1924. Tōbyō-mondō (*The Q&A of Fighting the Illness*).

Kido, Rie. 2007. "Towards the theorizing of 'tōjisha narratives.'" [*'Tōjisha no katari' no rironka ni mukete*]. *Sociologos*. No. 31: 86–98.

Kodaira, Tomoe, and Takehiko Ito. 2011. "When mental health patients are the primary actors: Using narratives by schizophrenia tōjisha as a clue (*'Tōjisha ga shujinkō' no jissen no arikata wo kangaeru: tōgōshicchoshō-tōjisha ni yoru naratibu wo tegakari ni*)." Wako University Sōgōbunka Kenkyūjo Nenpō, *Tozainanboku*, 130–140.

———. 2009. "Disability Memoirs as Narrative Teaching Material—Education through Storytelling by Mentally Ill People in Various Forms of Media (*Tayōna media ni okeru seishin shōgaisha no katari no kyōikuteki katsuy*)." *Macro Counseling Kenkyū* No. 8: 50–67.

———. 2008. "Mental Disability Memoirs—Multiple Meanings of Narratives (*Seishin shōgai no tōbyōki—tayōna monogatari no igi*)." *Macro Counseling Kenkyū* No. 7: 48–63.

Kunka, Andrew J. 2018. *Autobiographical Comics*. Bloomsbury Academic: New York, NY.

Køhlert, Frederik Byrn. 2019. *Serial Selves: Identity and Representation in Autobiographical Comics*. Rutgers University Press. New Brunswick: New Jersey.

Liggins, Jackie, Robin A. Kearns, and Peter J. Adams. 2012. "Using autoethnography to reclaim the 'place of healing' in mental illness." *Social Science & Medicine*, 91: 105–109.

Lindsay, Rachel. 2018. *RX: A Graphic Memoir*. New York: Grand Central Publishing.

Matsuda, Yasuko. 2018. "A critique of quality research on mental disability tōjisha experience (*Seishin shōgai tōjisha no keiken ni chakumokushita shituteki kenkyū ni kansuru hihanteki kentō*)." *The Bulletin of Research at Hokkaido University Graduate School, School of Education*. No. 132: 149–165.

Merrigan, Tara Wanda. 2018. "Girl, Interrupted, Twenty-Five Years Later." *The Paris Review*. June 28, 2018. https://www.theparisreview.org/blog/2018/06/27/girl-interrupted-twenty-five-years-later/.

Mullin, Chuck. 2019. *Bird Brain: Comics about Mental Health, Starring Pigeons*. Kansas City, MO: Andrews McMeel Publishing.

Myers, Kimberly R. and Michael D.F. Goldenberg. 2018. "Graphic Pathographies and the Ethical Practice of Person-Centered Medicine." *AMA Journal of Ethics*, 20(2): 158–166.

Nishikura, Miki. 2017. "Humor for Survival—Laughing at Oneself, Lightening Everyday Struggle. [*Ikiru tameno yūmoa—Jiko wo warau, yamai no nichijo wo warau*]." *The Horizon of Storytelling: Life Story Review. Journal of Japan Live Story Research Institute*, 2: 3–24.

Noguchi, Yuriko. 2018. "The examination of 'the life you want to live/the life you want to end' in Japanese illness narratives (*Nihon no tōbyōki ni okeru 'jibunrasii sei/jibun rashii shi' no arikatani tsuite no kousatsu: 'Tōbyō' ki eno henyō no kanten kara*)." *Shakai-shirin: Hosei Journal of Sociology and Social Sciences*, 64(4): 89–107.

Price, Margaret. 2011. *Mad at School: Rhetorics of Mental Disability and Academic Life*, University of Michigan Press.

Reaume, Geoffrey. 2017. "From the Perspectives of Mad People." In *The Routledge History of Madness and Mental Health*, edited by Greg Eghigian, pp. 277–296. New York: Routledge.

Rembis, Michael. 2018. "Review of Eghigian, Greg, ed. *The Routledge History of Madness and Mental Health*." *H-Disability, H-Net Reviews*, February, 2018: 1–4.

Sakae, Setsuko. 2016. *Seishin shōgai tōjisha no katari gamotarasu shakai henkaku no kanōsei*. [The Possibility of Social Change by the Public Talk of People with Mental Disorders]. *Core Ethics* No. 12, 89–101.

Sawada, Tadato. 2014. "Metaphoric dialogues as a technique—the sociology of tōjisha application and its 'inability to narrate'" (*Hiyuteki taiwa toiu giho—'katarienasa' wo meguru tōjisha jissen no shakaigaku*). *Mita Sociology* No. 19: 34–54.

Shields, Brooke. 2006. *Down Came the Rain: My Journey Through Postpartum Depression*. New York: Hachette Books.

Shigematsu, Kiyoshi. 2002. *Kiyoshiko*. Tokyo: Shinchosha.

Sklar, Howard. 2012. "Narrative Empowerment through Comics Storytelling: Facilitating the Life Stories of the Intellectually Disabled." *Storyworlds: A Journal of Narrative Studies* 4: 123–149.

Spiegelman, Art. 1986. *Maus: A Survivor's Tale*. Pantheon Books: New York, NY.

Stack, Frank, Joyce Brabner, and Harvey Pekar. 1994. *Our Cancer Year*. New York: Running Press.

Watson, Amy C., Patrick Corrigan, Jonathon E. Larson, Molly Sells. 2007. "Self-Stigma in People with Mental Illness." *Schizophrenia Bulletin* November 33 (6): 1312–1318.

Wilson, Ann and Peter Beresford. 2002. "Madness, Distress and Postmodernity: Putting the Record Straight." In *Disability/Postmodernity: Embodying Disability Theory*, edited by Mairian Corker and Tom Shakespeare, pp. 143–158. New York: Continuum.

Yagi, Gohei. 2009. *Learning about schizophrenia from first-person narratives: Returning to the starting point of psychiatry* (*Shuki kara manabu tōgōshicchōshō: seishin-igaku no genten ni kaeru*). Kanehara-shuppan: Tokyo, Japan.

CHAPTER 4

Essay Manga: Japan's Autobiographical Comics

The Definition of Essay Manga

What are essay manga? Essay manga are a genre of autobiographical comics that emerged and became popular in Japan in the 1980s. As the Japanese equivalent of autobiographical comics, it is the genre under which the majority of *tōjisha manga* titles are currently labeled. Essay manga are often described as "casual" readings of "serious" social issues. What distinguishes essay manga from traditional, word-only books are not just its accessibility but also its unique and exclusive ability to express things visually and metaphorically. These manga are comics based on the author's daily occurrences and real-life experiences, which are becoming increasingly popular in the world of social media (*Pixiv Essay* 2016). This genre has permeated consumers who do not regularly read manga (Takeuchi 2019) and has garnered so much demand that bookstores have begun setting up a special section just for essay manga titles. Subject matter varies from interracial marriage, dieting, language learning, personal finance, and parenting to medical treatment of serious illnesses such as depression (Narioka 2012). Unlike the conventional story manga in which the characters' perspectives are primarily provided via their speeches and actions, essay manga are designed to convey the author's perspectives, feelings, messages, or explanations as background information via first-person narration. Essay manga can be defined as comics that feature serious themes such as mental health issues and child-rearing difficulties and

Y. Okuyama, *Tōjisha Manga*,
https://doi.org/10.1007/978-3-031-00840-5_4

teach us the wisdom of life through the healing power of humor (Yamashita 2009). Essay comics are, Yamashita (2009) states, an example of Japanese literature that includes traditional diaries, essays, and novels, but is only expressed in the form of comics.

The word *essay* derives from the Japanese literature genre of essays (also called *zuihitsu*). The genre is also called *komikku essei* (comic essay) and *taiken essei* (experiential essay) interchangeably, although other terms such as *rupo manga* (reportage manga, especially for its investigative content), *watakushi manga* (I-manga), and *jitsuyō manga* (how-to manga) exist for minor use. Overall, *essay manga* and *comic essay* are used most frequently and interchangeably (Kawahama 2007). For consistency, the unitalicized spelling, essay manga, will be used in this book in the same way as the unitalicized tōjisha and *tōjisha manga*.

The Rise of Essay Manga

Although this genre has not been studied by Japanese comics studies experts as much as the other popular genres, several attempts of essay manga analysis have been made by Japanese comic analysts and academics. The prototype of essay manga first appeared in regular, story manga (*shōjo* and *shōnen* manga genres), according to Takeuchi Miho (2019), a manga studies specialist who analyzes popular manga.[1] Takeuchi describes the prototype's characteristics as follows: (1) the author is typically a female *mangaka* and usually appeared as a self-portrayed character at the beginning and/or the end of a *shōjo manga* (manga for young female readers) episode, (2) the character is drawn simply and acts comically, and (3) the character serves as a narrator of the story. Takeuchi interprets the character's role as a *baikai-sha* (go-between/intermediary agent) for the anonymous reader and the manga artist. The character does not always self-identify as the author, although, in many cases, it clearly represents the author, sharing distinctive physical features such as hairstyle, glasses, or a trademark hat. The purpose of inserting the author icon (or a look-alike character) into the story is to narrow the perceived distance between the reader and author, as though she exists in the same physical space as the reader. Takeuchi argues that the manga's status as a medium of popular culture enables the *baikai-sha* to make fun of herself with self-deprecating humor far more easily than other types of media such as film, successfully shortening the reader's sense of distance from the author. Takeuchi explains that the prototype characteristics of today's essay comic protagonists reside in this simply and comically crafted *baikai-sha* agent in the *shōjo* manga. However, it is important to note that these early *baikai-sha* episodes appeared only as "bonus" pages or an afterword to a *shōjo manga* comic book, not as entire stories, as they do now as essay manga.

Several scholars claim that Japan's genre of autobiographical comics was established when scores of self-narrative titles, in which comic artists used their own lives as subjects, began to be published in the 1980s.[2] The genre then proliferated between the 1990s and 2000s. For example, *Pixiv Essay* (2016) considers the early 1990s as the time when readers began to notice Saibara Rieko's *A Wanderer's Diary of Mahjong* (*Mājan hōrō-ki*) and popular comics of child-rearing as an emerging genre in the manga market. In the 2000s, Kurata Mayumi's *Loser-Men's Walker* (*Damenzu-Wuōkā* as a parody of a Japanese men's magazine, Men's Walker) in 2001, Oguri Saori's *My Darling is a Foreigner* (*Dārin wa Gaikokujin*), which began to be serialized in 2002, and several other highly popular graphic self-narratives continued. The boom was followed by similarly well-received works targeting "salarymen" (male salaried workers, more specifically, Japan's male, white-collar workers) such as Yoshitani's 2007 manga, *I, Otaku Salaryman* (*Boku, Otariiman*), and essay comics for cat lovers, such as Kuruneko Yamato's *Kuru Neko*, a blog-based series which began to be posted in 2006. *Pixiv Essay* refers to the most recent titles produced by Hosokawa Tenten and Ikeda Akiko as well.[3]

Takeuchi (2019) attributes the genre's birth to the time in which the female comic artist Sakura Momoko's *Chibi Maruko-chan* became a mega-hit manga in the 1980s. The *Chibi Maruko-chan* series was first published in a *shōjo manga* magazine by Shueisha in 1986 and was made into a popular TV show of the same name in Japan. Since her big breakthrough, there has been a proliferation of autobiographical comics. Thus, when the genre rapidly grew between the 1990s and 2000s, the majority of authors were still female manga artists. Although a more careful examination is needed to determine who the father or mother of essay manga is, Takeuchi sees Sakura Momoko as the *mangaka* who mainstreamed this experimental style of storytelling through *shōjo manga*. Therefore, while Takeuchi acknowledges the existence of prototype essay manga, she considers the late 1980s to be the time essay manga was established as a genre, crediting Sakura's publication of *Chibi Maruko-chan*. In that sense, Sakura Momoko was to Japan's essay manga as Justin Green was to US autobiographical comics.[4]

What made *Chibi Maruko-chan* stand out in the world of *shōjo manga* and establish itself as possibly the "first" essay manga? Takeuchi (2019) states the answer is in Sakura's drawing and storytelling styles that were never implemented before. First, the main character of the story, Maruko, was drawn as a non-descriptive, third-grade elementary school student with speckle-like eyes, unlike a typical *shōjo manga* heroine with eyes disproportionately large (often with sparkling stars in the pupil) while the nose is small, to look attractive and desirable. Sakura used no multilayer panel layouts nor the decoration of flowers and stars in the background drawings, both of which were common *shōjo manga* strategies. Instead, she kept the background simple and created characters, including the heroine, using a series of techniques called *deforume* (deformation). Second, Sakura's story consisted of episodes based on actual events, such as a universal experience of school life and her real-life family and neighbors. The episodes were permeated with self-deprecating jokes and perceptive (sometimes stinging) humor. Because of her comical characters and humorous plotline, her works were regarded also as *gyagu manga*, or gag manga (the genre of hyper-comedic comics using slapstick jokes and funny plots) even though the series was published in a mainstream *shōjo manga* magazine, *Ribon*. Furthermore, each episode is a self-contained story in which the narrator-protagonist Maruko-chan makes a funny yet astute comment on the occurrences from a third-person-observer's point of view.

The common characteristics of Sakura's works, represented by the long-lasting series of *Chibi Maruko-chan*, are heartwarming yet poignant observations of life. When a tale of ordinary people is told in intimate detail, it yields both realism and relatability. By examining herself from a bird's-eye view and laughing at her own goofiness, Sakura created self-effacing humor in her manga, and that tactic garnered a huge crowd of fans. The aforementioned characteristics of *Chibi Maruko-chan* series—relatability, self-deprecating humor, candid and poignant "third-person" observation of the author's personal experiences—clearly resonate in today's essay manga. Furthermore, these characteristics are also shared in the *tōjisha manga* titles featured in this book's analysis chapters.

Is it, then, accurate to designate Sakura Momoko as the "founder" of the essay manga genre? Kawahama (2007) found the oldest record of the term *essay manga* used in Fujiko Fujio's work, *Pāman no hibi* (*Daily Life of Parman*) published in the comic magazine Big Comic in 1978. That work documented Fujiko's personal anecdotes and insider stories surrounding his popular manga titles. It was advertised as "the first attempt to present an essay manga in the manga publishing industry!!" That record indicates that the prototype essay manga might be Fujiko's innovative endeavor made almost ten years earlier than *Chibi Maruko-chan*. One of the previous studies cited in Kawahama's paper is Kondo Masataka's (2006) analysis of Saibara Rieko's manga, which Kawahama praised as the most sophisticated examination of the genesis of essay manga. Kondo estimates that essay manga was established as a genre in the 1980s, similar to the theory of Takeuchi (2019). However, Kondo traces the prototype essay manga back to male gag manga author Tori Miki's *Ai no saka agari* (Gymnastics of Love) serialized from 1985 to 1986 in the magazine *Heibon Punch*. In that manga, the protagonist Mikio, wearing a pair of sunglasses, appears to be the author himself and visits various places in search of true love. Another researcher, Saida Takahiko (2001), also cited in Kawahama's paper, argues that *mangaka* such as Azuma Hideo and Ishikawa Jun, who drew avatars as their caricatured selves and wrote about their personal struggles, family troubles, and relationship issues, are in fact the "fathers" of essay manga, asserting that these *watakushi* manga (I-manga) led to today's essay manga.

As Kawahama (2007) states, despite minor differences, these scholars uniformly point to works published in the 1980s or slightly earlier as the emerging form of essay manga. It should be noted that none of these scholars are mentioned *Chibi Maruko-chan* as the original essay manga, however. Furthermore, in *Norakuro*, a pre-war gag manga by Tagawa Suiho, a figure resembling the author often appeared in the series. Similarly, in the 1970s, some artists invented a self-portrayed character who appeared as the narrator. Back then, the term *essay comics* did not yet exist, and the term *watakushi* (myself) manga or *shishōsetsu* (I-novel) manga were used instead. With these differing claims, it is not fruitful to trace the origin of this genre to a single work, nor is it useful to label any author as the "father" or "mother" of essay manga.

Connections with *Shōjo Manga*

It seems clear, however, that *shōjo manga* played a significant role in the development of essay manga. As Kawahama (2007) argues, there were experimental cases of essay manga precursors elsewhere. In *shōjo manga*, a one-page article called "Manga News" (*Manga dayori*) was issued beginning in 1970. A featured *mangaka* presented an episode about her daily life, and her news, published in the *Bessatsu Shōjo* Comics magazine, was to inform the artists' fans of her new manga title. The same magazine also began a newsletter "One Page Theatre" from 1971 to 1975, featuring popular *shōjo mangaka* such as Oshima Yumiko and Hagio Moto, and the newsletter published essays about their recent activities, such as Hagio's travel in England, which included illustrations of their self-portrayals. The *Bessatsu Shōjo* Comics published another monthly article by a *mangaka* whom the magazine sent to a place or event that comic fans suggested. In that publication, the *mangaka* reported their observation in a form of essay writing, which bears a close resemblance to today's *rupo* manga (reportage comics). Kawahama asserts that these experimental projects led to the proliferation of essay manga in the 1980s.

MOE's January 2011 issue contains an article titled "Comic Essay's Great Power," compiled by the magazine's editorial team, featuring popular essay manga titles.[5] The article provides a brief history of essay comics in three major phases: "the dawn of the new era" starting in the 1970s, "the growth period" in the second half of the 1980s, and "the Big Bang" of the twenty-first century. Representative cases of the first phase are Yamagishi Ryoko's *Ghost Tales—My Ghost Experience* (*Yureidan—Jibun no reitaiken*) in 1973 and Oshima Yumi's *My Toy Box* (*Yumin omochabako*) in 1978. The second phase includes Sakura Momoko's *Chibi Maruko-chan* in 1986, Saibara Rieko's *A Wanderer's Diary of Mahjong* in 1990, and Takahashi Yukari's *My Turkey Memoir* (*Toruko de watashi mo kangaeta*) in 1992. Among other examples, the last phase includes Kurata Mayumi's *Loser-Men's Walker* in 2001and Oguri Saori's *My Darling Is a Foreigner* in 2002.

After making the disclaimer that it is hard to determine whose manga work is the origin of essay comics, MOE's anonymous writer states that it is the beginning of the 1970s when short manga stories called *repōto* (reportage) manga began to be published, albeit on a minor scale, documenting the authors' own activities such as traveling, attending concerts, and other event participation in the *shōjo manga* arena. Around the same time, the aforementioned *mangaka* Yamagishi and Oshima also published longer non-fiction stories, which makes the MOE writer speculate that they might be the "pioneers" of essay comics. Similarly, in the 1970s, a minority of manga *tankōbon* (manga converted from a magazine series into a book) titles contained an *atogaki* (afterword, postscript) that would tell the reader about the author's current status, production secrets, and so on, which the writer thinks led to the development of today's essay manga. During the second phase, the most popular topics of essay manga were child-rearing and traveling.[6] Then, in 2005, *the Comic Essay Theater* debuted via the homepage of the magazine *Da Vinci*, where the term *comic essay* (essay manga) and the genre were publicly acknowledged. Therefore, it appears that several prototypes existed before the 2000s' "Big Bang" of essay manga boom. In other words, non-fiction graphic novels were published long before a label was established.

As speculated both by Kawahama (2007) and MOE (2011), essay manga probably emerged as an offshoot of the *shōjo* manga. Unfortunately, Japan's graphic memoirs were not classified into a specific genre until the 2000s, and only minimal research has been conducted by comics studies scholars to investigate what sociocultural factors contributed to the rise of essay manga, as opposed to I-novels which were already published in the early 1900s in the field of literature.[7] However, it is uncertain to what extent the 1900s' I-novel trend in literature influenced manga beyond the naming of *shishōsetsu manga* (I-novel manga) of the 1970s and why it took almost seventy years for self-narratives to be included in the field of comics. What is clear at this point is that although male manga artists somewhat contributed to the development of essay manga, the rise of essay manga is strongly connected to the *shōjo* manga renaissance of the 1970s in Japan.[8]

Characteristics of Essay Manga

The previous section on the rise of essay manga presented the prototype's characteristics such as a simply drawn protagonist who acts comically and plays the role of *baikai-sha* (go-between/intermediary agent) from Takeuchi (2019). This section provides the characteristics of contemporary essay manga discussed by other manga studies scholars.

In general, comics are a multimodal discourse consisting of visual and verbal signs. One of the important characteristics of essay manga is its being not only multimodal but also visually different from other manga (Narioka 2012). For example, essay manga typically uses verbal information visually—that is, how written text ("verbal signs") merges with image ("visual signs") in a way that story manga does not. In this regard, essay manga is a hybrid form of the two different media outlets of manga and essay writings, or a written essay presented in the form of comics. As a representative of essay manga, Narioka (2012) analyzes how Oguri Satari's mega-hit essay manga, *My Darling Is a Foreigner* (*Dārin wa gaikokujin*), hereafter *My Darling*, used various strategies to express visual information and meta discourse along with written text. One distinctive characteristic is essay manga's relatively high-frequency use of background texts (e.g. narrations, which are typically handwritten and placed outside of speech balloons).[9] Handwriting is a method considered effective for the essay manga author to convey an opinion in a casual or soft manner, as well as to personalize the messages as "reader-friendly." In this genre, key verbal text is the author's essay; thus, the frequent appearance of handwritten comments indicates the strong agency of the author. Furthermore, the line between the main character's speech and the author's narration often becomes blurry in essay manga because the two are essentially the same person. By contrast, multiple perspectives are provided by having other characters' speech bubbles appear alongside the author's narration within the same frame. Another important characteristic of essay manga is that characters, including the author herself, are drawn simply and comically, as exemplified by Oguri Satari's *My Darling* (Narioka 2012). This feature was also listed by Takeuchi (2019) in analyzing Sakura Momoko's (1987) *Chibi Maruko.*

To summarize Narioka's analysis, the characteristics of essay comics are as follows: the author as the protagonist of a story; the story based on his or her own real life; heavy use of background text, especially handwritten, carefully explained, first-person narrative text; vivid, often exaggerated facial expressions for humor and entertainment effect; and simply yet comically drawn characters. To Narioka (2012), these characteristics cleverly produce the effects of friendliness and closeness in essay manga. It is important to clarify that, as Narioka pointed out, characteristics such as humorous simplicity typically seen in essay manga are not the author's personal style but are the genre-specific characteristics.[10]

Possibly, Natsume (2006) was the first person who attempted to analyze essay manga in his popular book, *What's Wrong with Learning About Life from Manga?*, when scholars of manga studies rarely paid attention to that budding genre. In that book, Natsume praises Morimoto Kozueko's *I'm Your Mom* (*Watashi ga mama yo*), a comic memoir about Morimoto's own parenting, published in 1996 as one of the earliest works to exemplify that the genre of *shōjo manga* "had matured to produce essay-like narrative of literature such as this work by the end of the late 1980s" (207). It is important to point out that at the time Natsume's (2006) book was published, the label *essay manga* was not widely used, nor was the genre itself clearly defined. Thus, self-narrative gag manga such as Morimoto's and Hara Keiko's were called "essay-like" manga in his book.

Regarding the characteristics of this "mature," "essay-like" manga, Natsume (2006) describes various strategies developed by manga artists to visually depict subtle emotional states of characters, which he first noticed in manga titles published in *josē* manga, the genre for adult female comic readers.[11] In his view, the manga that deal with important issues germane to women such as parenting, romance, sex, and marriage, and in some cases, even larger social themes involving women, are considered "mature." Natsume also discerns the "*kawaii*" (cute) illustration of characters, including anthropomorphism ("*gijin kyara*"), as another trait of "essay-like" manga.[12] In one example, Uchida Shungiku's *We're Procreating* (*Watashitachi wa hanshoku shiteiru*), a 1994 comic based on Uchida's own parenting experience, Natsume argues that this type of manga is no longer just for entertainment; it teaches the reader about life's many challenges and hurdles (247). Here, he stresses that many other manga studies scholars with whom he worked in Japan also shared the positive sentiment toward "essay-like" manga. Therefore, by the early 2000s, this genre was already known as a group of manga from which readers gain valuable insights into how the protagonists tackle their personal issues with courage and, importantly, with a good sense of humor. Such "lessons" can include how to reverse the course of *ikizurasa* (difficulty of living) associated with mental health.

MOE's 2011 article is also informative about essay comic characteristics through their analysis of representative works. In the anonymous writer's words, essay comics are a perfect choice "when readers do not feel like reading comics of large size" and have produced many new talents for this niche market. Various essay manga are recommended for several distinctive categories. MOE (2011) selects Hosokawa's *My Significant Other Has Got Depression* as an essay manga title for the subject of "*shimi-jimi shitai*" (pathos in others' lives). The titles MOE's article listed under this category of pathos are commended for their excellence at addressing issues for mature readers, stirring up emotions of sympathy and sorrow in readers, and leaving them with tears and warmth at the same time. Besides Hosokawa's manga, all the other *tōjisha manga* titles I analyzed for this book also display this characteristic. (See more on this matter in the next section)

MOE's (2011) other categories of essay manga include "how-to" (e.g., Ikeda Kyoko's *How to Structure Your Day: Productivity Coaching*, or

Ichinichi ga miete raku ni naru, jikan seiri-jutsu), "professions" (works describing a certain career), and "animals" (mainly about pets). Because essay manga stories are based on authors' real lives, readers are more likely to feel close to the manga artists. Two other popular categories of essay manga are "slice-of-life" and "tryouts (*taiatari-kei*)." The "slice-of-life" category refers to a type of story in which the writer appears as a comical boy- or girl-next-door type of character who finds humor and happiness in ordinary, yet relatable things such as grocery shopping as in Kera Keiko's *Shopping Together* (*Issho ni sūpā*). The tryouts type, exemplified by many of Saibara Rieko's comic essay titles, hallmarks the protagonist who goes anywhere and tries out anything just for the sake of experience. Below is an excerpt (my translation) of the MOE interview with Saibara:

> At first, I conceptualized this genre as a manga version of *sanbun* (informal prose) similar to travel writing by a novelist. I adopted a tough middle-aged woman's attitude (*obachan seishin*), which is one of the secrets of my success in investigating thoroughly and collecting hard facts, no matter how difficult the investigation becomes. I then present the information mixed with severe "straight-face joke" (*sukkomi*) and "shticks" (*gyagu*) to entertain the reader. To me, the key to the comics' popularity is my tireless effort of buttering up (*omoneru*) readers. I believe manga should be entertaining. I achieve that goal by showing my toilsome-self and generating laughter from the reader.

Incidentally, Saibara's very first essay manga is *Forgetful Travelogue*, or *Tori-atama kikō*, a reportage about her traveling experience punctuated with insightful yet satirical observations and useful information. Thus, it makes sense that her initial conceptualization of essay manga was travel writing. Saibara confesses, however, that recently, writing shtick-filled manga has worn her out. Thus, she began to work on picture books such as *Your God, My God* (*Kimi no kamisama*) simply to "rehabilitate" herself. The shtick strategies of *tsukkomi* (provocative gibes) and *gyagu* (gags, especially slapstick jokes) are often mentioned by other essay manga artists, and I will elaborate on that issue in the next section.

Other significant comic essay characteristics also emerged from the interviews conducted by the writers of MOE (2011). For example, in creating a self-narrative, the essay comic artist takes copious notes as they observe the world and adds simple illustrations to the notes. They also try to keep the essay "light and fluffy" (*hunwari*) and the characters "realistic yet with some deformation of drawings" (*deforume shiteru kedo, riaru-kan ga aru*). While some essay manga attempt to entertain through the story's ambience, others use explicit punchlines to do the job. Getting back to the category of "tryouts," its key characteristic is that the story is conveyed with a sense of reality so readers feel as if they are vicariously experiencing the tryout, whether it is running a marathon or living alone for the first time. The article emphasizes that regardless of the categories, "empathy buying" (*kyōkan-gai*) is a commercially important element of essay comics. First, empathy is formed through a similar living situation. For example, the reader might relate to the author immediately if both of them share the same marital situation or pet ownership. Second, the degree of empathy is also influenced by age. Third, readers' needs are equally important as their interest in manga increases if it offers a simulation effect when taking up a certain profession or a type of healing, empowerment, or learning that the reader desires.

Komikku Essei Gekijō (Comic Essay Theatre), a website-based editorial department of Kadokawa Corporation (Kadokawa Shoten), publishes essay manga titles for "SNS" outlets and runs the essay manga "Rookie of the Year Award" competition for budding manga artists. (SNS stands for Social Networking Service and is a commonly used acronym in Japan, meaning collective social media sites or channels such as Twitter, Facebook, Instagram, LINE, and MachiMachi.) Its official website, *Web Komikku Essei Gekijō*, defines the genre of essay manga as "comics that feature the author's lived experience unlike regular manga" (https://www.comic-essay.com/info/), in which the essay manga is contrasted with the *futsū no* (regular) manga, referring to traditional *story manga*. The site page also characterizes essay manga as "works that make you empathize with, laugh, and sometimes sob." The essay manga released by this department are mostly nonfictional (e.g., Nohara Hiroko's *May I File for Divorce?*). However, there are a few exceptions of semi-autobiographical comics that mix real life with fantasy (e.g., Sakura Tamakichi's *Mandama Diary*). Major essay comics sponsored by Kadokawa Corporation include works created by established manga artists: Hasegawa Machiko's (2016) *Sazaesan Uchiakebanashi*, Fujiko Fujio's *Manga-michi*, Mizuki Shigeru's *Comic Showa History*, Sakura Momoko's *Chibi Maruko*, Saibara Rieko's (2004) *Mainichi Kāsan*, Hanawa Kazuichi's (2018) *Keimusho no naka*, Azuma Hideo's *Shissō Nikki*, and Ogura Satari's *My Darling Is a Foreigner*. Whether the comic is a fully or semi-autobiographical, essay manga has distinctive styles, they claim. These styles are described by Yamazaki Shun, the Associate Editor of the department, as follows: Our essay manga titles depict the author's deep thoughts and opinions on life events like essays and are, unlike ordinary manga, illustrated mostly with simple drawings and the so-called *datsuryoku-kei*, or comical style; its characters are typically drawn to be funny and cute dollish figures (a type known to attract many fans). He claims that many professional manga artists had a debut from this site and that an increasing number of newbies have attempted to write about their unique experiences in essay manga. Therefore, Yamazaki believes this genre has huge growth potential in the future.

A 2016 essay posted on the homepage of *Pixiv Essay*, a web-based magazine that showcases newly released essay manga titles, boasts that many of these works end up getting published in book form and become bestsellers. This web comic magazine was created as a collaboration between Pixiv and KADOKAWA's editorial department of essay manga. Comics originally published on the Internet (e.g., works posted in blogs, Pixiv, Twitter, or Instagram) occupy a large part of the essay manga genre in Japan. According to the chief of staff Ishii Shintaro, SNS and essay manga are made for each other. Yamazaki at KADOKAWA further explains the genre's compatibility with the Internet as follows: Because the content of the essay manga covers the author's own struggles and surroundings, if the story is relatable and yields high affinity, word goes around very quickly on the Internet. Due to the widespread use of smartphones in Japan, the number of SNS users has been rising these days. If the work is favored among SNS users, essay manga is very likely to become a bestseller via this venue.

It should be reiterated that essay manga authors do not disclose everything about their struggles and surroundings; they decide what parts of their experiences to put into print. Some manga artists add fictional elements to protect the privacy of the real people who appear as characters in their stories (e.g., Uchida Shungiku's *We're Procreating*). Privacy protection, not only for the artists themselves but also for their families and friends, is an important consideration, particularly if the theme of the manga is a mental disability. In fact, one of the manga artists I interviewed raised this concern. Interestingly, some *mangaka* disclose in their works that the protagonist is not completely identical to themselves but is a figure that the artist "has sublimed (or lionized)" (*shōka saseta*) for publication. Regardless, the core story is autobiographical, which separates this genre from the rest.

Pixiv Essay (2016) asserts that the authors must meet the following three prerequisites if they choose to publish their essay manga in digital format and want to attract smartphone readers quickly: affinity, a carefree attitude, and a small volume for reading in small pockets of time. In fact, the requirement of carefree reading is frequently suggested as an appeal of this genre regardless of the publication format. Moreover, essay manga's resemblance to a personal diary attracts even people who do not draw manga on a regular basis, producing a surplus of budding manga artists and occasional mega hits in the manga market. On the homepage of *Pixiv Essay,* the editor Yamazaki admits that essay comics are not as widely featured as regular manga titles in the mass media. Yet, he emphasizes that the genre produced titles of which more than 100,000 copies have been sold and adds that websites that release new titles for free also attract many fans. In the manga market, 100,000 copies are considered the figure that separates "a hit" comic from the rest (Takemiya and Uchida 2014). Thus, Yamazaki's number legitimizes essay manga's strong presence within the manga publishing industry. He also lists the following characteristics of essay manga based on his years of experience with that genre:

1. The initial target readership of essay manga was 20-to 30-year-old women, but now the number of male readers from the same age group is increasing, especially with businessmen for whom a special niche market exists.
2. Its popular themes range from child-rearing to LGBT topics to non-traditional hobbies (e.g., mountain climbing for women; being a railroad enthusiast) and highly specialized occupations such as an officer at the Japan Self-Defense Forces. Because essay manga's initial readers were women, topics such as romance, traveling, parenting, and relationships continue to be in high demand, although so-called male-oriented professional jobs, skills, hobbies, and activities are being featured in recent essay manga in response to increasing male readership.
3. Unlike regular manga's typical protagonist who is a sexy, good-looking character, essay manga's hero or heroine tends to be a mocked, merry characterization of the author's self.
4. Essay manga's artistic style is simple. It does not use a complicated panel layout, either. Its simplicity has a two-way benefit: It allows

> rookie artists to draw manga effortlessly by mimicking others, and its undemanding illustration works well with inexperienced readers since they can enjoy the work without knowing much about manga-specific conventions.

Chu and Coffey (2015) analyze Takagi Naoko's *Hitoritabi ichinensei*, which was published in 2006. According to their research, this essay manga became quite popular not only in Japan but also across Southeast Asia, selling more than 250,000 copies in that region by 2012. In the manga, Takagi's self-portrait appears to these scholars like "a little girl without a trace of womanly characteristics" (152). Indeed, most heroines of essay manga are drawn with a simple face wearing little or no makeup and a flat body with no narrow waist nor clear bust and hips. They attribute the drawing of such an image to "the influence of the kidult phenomenon (20-to 35-year old adults indulging in children's or teenagers' entertainment and hobbies)" and the Japanese "kawaii" culture in which "many adults consciously maintain some childhood behavior and appearance" (152). It might appear puzzling at first glance for the accomplished, solo-traveler heroine with her independent spirit to be drawn to such physical vulnerability. But such an undemanding artistic style is, as Yamazaki described in *Pixiv Essay* (2016), a characteristic and strength of essay manga. In my view, what is distinctive about the genre is precisely this imbalance between the heroine's visual vulnerability and her amazing feat, such as conquering the fear of traveling alone, surviving a rare disease, or tackling a mental health issue.

Another key characteristic Takeuchi (2019) finds is that in most essay manga, each episode is usually self-contained. Similar to the previous studies, he also distinguishes essay manga's unornamented illustration of characters from the more elaborate design of the standard story manga. Essay manga's simplicity is reminiscent of the style of *heta-uma* manga, Japan's underground manga movement in the 1970s. The word *heta-uma* literally means "poor but good," referring to the type of drawing that is seemingly poor but has an aesthetic charm, sharply contrastive with the polished drawing style of mainstream manga. Artists who drew *heta-uma* manga were capable of more elaborate drawing but chose the simplified image for its novelty and strong impact. However, essay manga's unpretentious characters are not identical with those of *heta-uma*, as cuteness and healing humor are the genre's other distinctive characteristics. In the majority of cases, the protagonist of essay manga is adorable, akin to recently prevalent *yuru-kyara* ("loose character") mascots in domestic tourism (Tan 2014) as well as *iyashi-kei* (mentally soothing type) merchandise goods (Matsui 2011) in Japan. The majority of essay manga characters are what we call *santō-shin* (literally, "three-heads tall," referring to a figure whose head is about one-third of the body). Unlike story manga's protagonists who are often depicted as sexualized objects, essay manga's self-representations are unembellished yet lovable, and quite often, they are exaggerated, lampooned avatars of the authors themselves, much closer to the characters of gag manga. In some cases, artists use cute animals or creatures as self-portrayals (e.g., a light blue balloon character of Tomoyo's (2018) *Watashi chūgakusei kara tōgōshicchō yattemasu*; a bipedal, bespectacled cow of Arakawa Hiromu's (2009) *Hyakushō kizoku*; and a Pillsbury Doughboy-like figure of Isayama Motoko's (2018) *Atai bijinbyōni narimashita*).

The last, but not least, characteristic of essay manga, which none of the cited previous studies has brought to light is its typical self-introduction at the beginning. For example, Arakawa's *Hyakushō kizoku* begins with a scene of a tea house with her editor, in which Arakawa's anthropomorphic character of a cow tells the reader that she is a *mangaka* who used to be a mud-covered, farmhand in Hokkaido. Likewise, Sato Miyuki's (2013) *Nanbyō nikki* about her battle with Adult-Onset Still's Disease (ASD) features the author herself saying "Nice to meet you! I'm Sato Miyuki. You probably didn't expect this, but I'm about to die" in the very first panel of the manga. All the *tōjisha manga* titles selected for this book also begin with this canonical self-introduction. This must be part of a strategy for relatability or affinity-building which the author elects to use in essay manga. Especially when the narrative deals with a sensitive subject such as mental health crisis, it is important to bond with readers and gain their trust.

Characteristics of *Tōjisha Manga*

A new development within the genre of essay manga has been noticed by academics who evaluate comics as a resource for mental health education (e.g., Kodaira and Ito 2008; Nishikura 2017): the insurgence of disability memoirs by manga artists who are either tōjisha with mental disorders or tōjisha caregivers. In this section, I will focus on this particular subgenre of essay manga and delineate the characteristics of "manga by tōjisha," or what I call *tōjisha manga*.

In 2006, manga artist Azuma Hideo (2005a) wrote a magazine article titled "I Resurrected by Writing a True-Life Story," in which he described how he turned his experience of alcoholism and homelessness into an award-winning comic story. After the astronomical sales of 170,000 copies, Azuma, who suffers from alcoholism, bipolar disorder, and anxiety disorders, was awarded the 2005 Grand Prize from the *Nihon Mangaka Kyōkai* (Japan Association of Manga Artists) for his *Shissō Nikki* (*Diary of Vanishing Self*) published by East Press. In the story, he vanishes one day, becomes homeless, and even attempts suicide; he suffers from severe hallucinations, hearing his own voice telling himself that he will become insane, develops unsettling feelings such as being ambushed on the street, and eventually, becomes hospitalized. The author reveals in the article that despite the fact that he openly wrote about some of his most despicable experiences, because his domestic affairs would be even more painful to talk about, he focused on his own circumstances, keeping his family from appearing in the story. This revelation indicates that a manga artist chooses when and what part of the experience to use for the public forum such as manga, instead of disclosing everything indiscriminately. He also states that the depiction of family matters would make gag manga "too serious," thereby losing its entertaining quality. In fact, Azuma could not create a story out of the whole experience of vanishing at first because it felt too vivid and unpleasant to reflect upon. In the end, he decided to write about it because he wanted to turn over a new leaf. Although further research is needed to confirm, Azuma's *Shissō Nikki* might be the first case of *tōjisha manga*.

One of the most frequently analyzed autobiographic comics on the topic of mental disability is a popular essay manga created by Hosokawa Tenten, *My Significant Other Has Got Depression* (*Tsure ga utsuni narimashite*), a trilogy published by Gentōsha between 2006 and 2013. The main character, Husband, is a character drawn in a style called *datsuryokukei* (relief and slice-of-life comedy). Because of Hosokawa's clever use of humor, the reader does not feel stressed by reading about a horrible experience of depression in her manga (Yamashita 2009). Ōhara Yukiko's *Mr. Ōhara and His Recent Neurosis* (*Ōhara-san chi no danna sann konogoro sukoshi shinkeishō*), also published in 2009, depicts Ohara's direct experience of living under the influence of her husband's panic disorders. Her husband, Ohara Kōki, published a non-comic memoir in which he describes his OCD-driven fears and treatment. A third example of often cited essay manga is Nakamura Yuki's *Wagaya no haha wa byōkidesu* (*My

Mother Is a Sicko), a vivid memoir of living with her mother with schizophrenia, published in 2008. All three were examined in Yamashita's (2009) paper.

Kodaira and Ito (2008) listed several examples of manga that feature the author's lived experience of psychological disability: works by Hosokawa Tenten, Azuma Hideo, Ōhara Kōki and Fujiomi Shūko, Tanaka Miru, and Nakamura Yuki.[13] The common denominator of these manga is a humorous and cutely drawn protagonist narrating seriously and realistically what it is like to have mental disability. Hosokawa's *My Significant Other Has Got Depression* was featured as an empathy-evoking comic story told in a carefree manner in *Asahi Newspaper* in 2006, the same year Azuma's disability memoir was published. It is important to stress that, along with the element of empathy, these manga also take a jolly, lighthearted approach to depicting their challenging days of misery, struggle, and frustration caused by their own or their family member's mental health issues. To get to the heart of the matter, these features are what characterize *tōjisha manga*. However, it is equally important to emphasize that when the tōjisha creates a manga story, most likely the author has reached the state of *kankai* (phase of stability). Because *tōjisha manga* authors are already "over the hump," so to speak, they are more likely to ease off compared to the tōjisha who is in the midst of fighting their mental health crisis.

Another characteristic of *tōjisha manga* is the "educational" role the manga plays. This subgenre of essay manga is said to have a great potential to improve the public image of mental disorders by taking advantage of the manga's inherent reader-friendly mode of expression (e.g., Kodaira and Ito 2008; Kubota 2006). According to Kodaira and Ito (2008), the nature of reader-friendliness derives from several facts: manga's appeal to a wide range of audiences, its high market penetration, and its accessibility. To elaborate on the merit of accessibility, they explain that those who do not normally read books about mental disability may be more willing to read about the topic in manga format. Consequently, the more people read such manga, the more information on mental disability is disseminated to the general public. Thus, they speculate that manga is more likely than other media to reduce prejudice against mental disorders in Japan.

Azuma Hideo's *Shissō Nikki* is said to be one of the earliest comics on the ins and outs of alcoholism, including his hospitalization and participation in Alcoholics Anonymous (Kodaira and Ito 2008). As mentioned earlier, the artist won a Tezuka Osamu Culture Award in 2006 with this work. Azuma (2006) wrote *Utsu Utsu Hideo Nikki* (*Melancholy Hideo Diary*), which describes not only his depression bouts but also his experience with psychopharmaceuticals and mental health clinics in his town. Together with Azuma's manga, Kodaira and Ito (2008) recommend *Seishinka ni iko!* (*Let's Go to a Mental Health Clinic!*), a two-volume graphic memoir about panic attack disorders written by Ōhara Kōki and illustrated by Fujiomi Shūko (2002), as a manga for educational purposes. The first volume of their manga series was published in 1999. Noticeably, the manga illustrates how reluctant people generally feel about visiting a mental health clinic for the first time, how anxiety-provoking the first visit can be, but how pleasant and smooth the experience actually is at the clinic. In its first episode, Fujiomi's self-character utters, "Did you know that a mental clinic is rather a peaceful place with ordinary clients, including patients just with physical illnesses" (5). According to Kodaira and Ito, Fujiomi was reported to have commented that she hopes mental health care facilities are more approachable and she published the book to encourage those who are too afraid of visiting such a clinic (1999 *Asahi Newspaper*).[14]

Kodaira and Ito (2008) introduce another *tōjisha manga*: Tanaka Miru's memoir about her bipolar disorder and BPD (Borderline Personality Disorder), published by Seiwa Shoten, a publisher of psychology and other medical studies books. This 2006 manga has a long title, *Manga About BPD and Bipolar with a Message for Those Struggling Daily: Go at Your Own Pace!* (*Manga kyōkaisei shinkaku shōgai & sōutsubyo remix: hibi funtōshiteiru kata he, mai pēsu de ikō!*). Analyzing Tanaka's work, Kodaira and Ito present another common characteristic of disability memoir manga: a physician's commentaries within the manga story. Similar to Fujimori's Dr. Ono, Tanaka's attending psychiatrist writes a short essay at the end of the manga, commenting on her work. This is also the pattern quite common in many of the *tōjisha manga* I reviewed. Kodaira and Ito also argue that graphic disability memoirs are excellent at illustrating medical experience, including how the tōjisha responds to certain symptoms. That is, they can express the internal state of the tōjisha visually from their first-person point of view, showing their otherwise invisible emotion through vivid and creative portrayals of their depression or other psychological conditions. Therefore, in Kodaira and Ito's opinion, Tanaka's manga is the narrative that offers insights into the tōjisha's perspective; it explains to the reader why the BPD patient often chooses an extreme action, for example. In this manga Tanaka also revealed that she is an

Adult Children of Alcoholics (ACOA), illustrating how she used to cry frantically, have panic attacks, feel flustered, and harm herself, which might be painfully realistic to some readers. In their 2009 paper, Kodaira and Ito praise Tanaka's work also for providing in-depth information such as diagnosis criteria and symptoms of ACOA in a comprehensible manner.

Kodaira and Ito (2008) strongly recommend this type of essay manga titles be introduced as textbooks in the field of medical education. They equate the status of essay manga with that of other educational outlets, including the Bethel House's website, "Bethel Tōjisha Kenkyū Room (*Beteru tōjisha kenkyū no heya*)"; Sōda Kazuhiko's 2008 documentary film, *Mental* (*Seishin*); and two television programs that air voices of schizophrenia tōjisha, "Live Brilliantly (*Kiratto Ikiru*)" and "Life, Relief, and Morning (*Seikatsu hotto mōningu*)" broadcast by NHK (*Nippon Hōsō Kyōkai*, or Japan Broadcasting Corporation). They also report that a stigma reduction program, developed in 2002 by the World Psychiatric Association (WPA), an international umbrella organization for psychiatric associations in various countries, acknowledged the effectiveness of incorporating manga as an educational tool for prejudice elimination. In that WPA report, essay manga was touted as a reader-friendly medium that attracts a wide range of people in a variety of social statuses. For these reasons, Kodaira and Ito encourage Japan's medical schools to use *tōbyōki* (illness memoirs, discussed in chapter "Tōjisha Narratives"), including mental disability essay manga, for stigma reduction education. Kodaira Tomoe herself uses *tōbyōki* texts regularly in her course on mental healthcare nursing studies. The course's student feedback indicates that it helped lessen their worries about becoming a nurse and assisted them in building their career identity. Except for the skills learned only from hands-on training, tōjisha narratives, including essay manga by tōjisha (thus, "tōjisha manga"), are great educational resources for multiple applications.

Kubota (2006) also recommends Azuma Hideo's *Shissō Nikki* as a good example of graphic memoirs featuring mental health issues. The manga documents Azuma's experience as an alcoholic and homeless man. His alcoholism led to hallucinations (a known alcohol-induced psychosis), and he also had panic attacks and suicidal ideation. His alcohol-induced psychosis were visual hallucinations, from which Kubota speculates that Azuma had an easier time expressing his symptoms graphically. As he explains, alcoholism is a serious mental issue because anyone can become alcohol-dependent if they abuse it for an extended period, and once they become addicted, their body never returns to its normal state. Azuma also suffered from depression and splendidly portrayed his mood swings from *sō* (being manic) to *utsu* (being depressed). People with Bipolar II experience only a hypomanic episode (no severe manic episodes like Bipolar I), yet they do have major (meaning "strong") depression episodes, during which, as Kubota describes, their mental state is akin to black worms creeping up in the head. To avoid those unpleasant feelings, many depressed people turn to alcohol, he explains.

Another *tōjisha manga* Kubota (2006) highly recommends is Tanaka Miru's *Easy-to-Read Comic About Bipolar Disorder's Ups and Downs* (*Manga otegaru sōutsubyō kōza High & Low*), published in 2004 by Seiwa Shoten because of its accurate and comprehensible depiction of bipolar disorder. (Tanaka is also the author of *Manga About BPD and Bipolar* discussed earlier.) Kubota, who is a faculty member at Shiga University and one of the counselors at the school's health maintenance center for students, appears to offer a medically reliable assessment in terms of manga's accuracy in illustrating psychological symptoms. However, I would like to point out that manga is not the medium that restricts artists to inventively depict the tōjisha's mental state. In fact, making the invisible visible is where manga excels. I will demonstrate how the selected *tōjisha manga* titles exemplify this capability of comics in each case study chapter. Here, I should also disclose that Kodaira and Ito (2008) reported the WPA's comment on comics' shortcomings: It is not easy for comics to embed serious messages in their carefree stories, and comics' culture-specific conventions and artistic peculiarities may not necessarily make manga a completely accessible medium for readers.

An Example of In-Depth Analysis

In this section, I will present a careful analysis of an autobiographical comic as a case of how an essay manga that depicts lived experience of illness can be examined in depth although her disability is not in the category of mental health. Nishikura's (2017) analysis shows how one manga artist, Komame, cleverly maneuvered a fine line between the medium's lightheartedness and the weight of a sobering topic. I also intend to present this case study as a way to segue into humor, an important topic of disability memoirs, in the section that follows.

Nishikura (2017) interviewed a 40-year-old manga artist, Komame Daruma, who suffered from alopecia areata, also known as spot baldness, which has no known cure. The interview was centered around the manga artist's two graphic memoirs: *A Beginner's Guide—Fight! Alopecia Areata* (*Watashi, pika pika no ichinensei—Dato! Enkei datsumōshō*) published in 2012 by Kadokawa Corporation and *Daily Hair Battle: 24-Hour Close Report of a Wig Life!!* (*Hibi zurezure: Katsura seikatsu micchaku 24-ji!!*) in 2016 by Shōgakukan. Komame is married to a citizen of Myanmar (then-called Burma) and also created another autobiographical manga in 2013, *My Husband Is a Burmese: Secrets of the World's Happiest Country* (*Danna sama wa būtan jin: Sekai-ichi kōfukuna kuni no himitsu*). Komame had chosen to be a gag manga artist because of her natural inclination toward laughing and funny things. Her hair loss started when she was a senior in college. By age thirty-four, her hair was completely gone. She has been wearing a wig ever since age thirty-five. At first, she could not accept her altered appearance and locked herself indoors. However, she was inspired by the life of her stalwart family doctor and decided to put her artistic skills to good use for others by writing about her experience of the disease. She chose to disclose it publicly by publishing her story in an entertainment magazine because she felt that the only way to solve social issues about this disease was by educating the general public about *alopecia areata*. However, realizing that her experience does not necessarily relate to other tōjisha's suffering, Komame later switched her focus from the particular diagnosis of *alopecia areata to more relatable topics such as* relationship issues due to her "being different" in a collective society.

What Nishikura found in her interview with Komame is how this gag manga artist strategized when writing her *tōjisha manga*: First, Komame separated herself as the *mangaka* writing the memoir from her tōjisha-self that is being written about in the story. This way, the former self is allowed to reflect upon the latter self in a creation process. Second, she figured out what would be okay and what would be too hard to illustrate within the drawing format of gag manga, which means that she was aware of the merits and limitations of her genre. Third, she incorporated the perspectives of autobiography to turn her life story into an essay manga. She chose to produce this as an autobiography in gag manga style because she wanted to find both hope and humor in her daunting experience of the disease. In doing so, she succeeded in finding some answers to the questions she had in mind and turned the painful experience into a positive outcome. By playing the object of laughter, she became more hopeful and developed a softened view of the matter. At first, what bothered her the most was her inability to find any humor in suffering from the disease. However, once she changed her point of view, she realized how comical her experience could be. To her, that realization was a gift. She successfully kept a psychological distance from that horrible experience and received payback by making the manga out of it, finding strength and boldness to power herself to fight the illness. In other words, the merits and demerits of openly disclosing her disease have evened out.

As a professional manga artist, Komame knew deep down that it could have turned into a painful work, had she taken a really serious approach to the issue. She was aware that telling her miserable experience might unnecessarily depress many readers and probably alienate those who are not tōjisha. She also sensed that she should not use everything that she experienced but must select which parts of her reality to craft into the story for what purposes. Komame emphasized the importance for professional artists to prioritize how the work is received publicly. She also stressed that some tōjisha do not want to be the butt of jokes at all. For this reason, she was initially a bit worried about using the gag manga format whose sole purpose is to make the reader laugh. But she also thought of other readers who may feel inspired by her story and find hope, so she decided to execute the plan. In addition, she felt the project would become her lifeline as well. The project also taught her to take an objective view of herself and pay attention to any comical aspects of her life. Through this process, she learned to have a bird's-eye view of herself and was able to objectify and sublimate her painful experience into a work of art.

Komame attributes her series' success to having herself represented by an "avatar" under the pen name, Komame Daruma.[15] In this autopathography, she appeared as a character named "Komame," not as her true self. In her words, the character is her other self, a "half-stranger." That mindset freed her from being self-conscious and allowed her to play the role of a creator-artist who reports on that character in the narrative, which is a process of her sublimation. Here, Nishikura makes an expository comment: according to Asano Tomohiko (2001), self-narratives entail a curious phenomenon of producing both the recounting "self" and the recounted "self." Positioning the self in two separate places—the present and the past—yields the "story realm" (*sutōrī ryōiki*) to which the narrator belongs and the "fable realm" (*monogatari ryōiki*) in which the protagonist (the author as the character) performs in the narrative. Thus, based on Asano's theory, Komame "duplicated" herself, one belonging to the story realm and the other in the fable realm, wisely selected "safe" events from her past, and created the hilarious autobiography of battling the disease. Here, it is important to highlight that in addition to the aforementioned self-duplication, she took the unique approach of distancing herself from the disease through humor. Thus, her manga amusingly narrates how she came to accept the reality of "no known cure" and survived both child birth and parenting while wearing a wig.

However, Komame also knows of the downsides of humor. As mentioned earlier, some tōjisha get upset when their condition is the object of laughter. Although that is a legitimate concern, the public understands that gag manga is a vehicle that allows authors to make fun of their own confusion and comical reactions to the onset of the sickness. If humor plays an antidote to the pain and enables her to tell her hardship with ease, then it is worth it, she thought. Her ability to laugh at herself is an advantage as the gag manga artist; she was able to find the right moment to *tsukkomi wo ireru* (provoke laughter) as a third-party observer.[16] To her, the "third-party" observation means detaching oneself from the matter and avoiding the sheer mental absorption of tōjisha life. I believe this is exactly what makes *tōjisha manga* special: tōjisha narrative allows for the separation between the *mangaka* self and the tōjisha self and enables the narrator to jab at her disability experience to provoke laughter. I will discuss in more detail the healing role of humor in disability memoirs in the next section.

The concept of *tsukkomi wo ireru*, or "provoking laughter," mentioned by Komame needs a careful examination here because it is one of the special qualities of *tōjisha manga*, as mentioned above. The word *tsukkomi* can be translated as "poking one's nose into something" or "digging deep into the subject." Nishikura discusses the comedic pair of "*tsukkomi* and *boke*," which is ubiquitous in Japanese culture. In that style, one comedian plays the role of *tsukkomi* by digging deeper into something the partner *boke* said nonchalantly. If the *tsukkomi* comedian effectively utters something hilarious and poignant, the audience bursts into laughter. For this comedic style to work well, two separate parties—the provoker (*tsukkomi*) and the provoked (*boke*) must perform in tandem. Applied in the context of her gag manga-styled autobiographical comics, the author plays the role of *tsukkomi* when narrating the tōjisha herself, who kicks and shouts in panic while combatting her disease. Nishikura emphasizes the important role of *tsukkomi* in Komame's manga because it is the essence that makes her disability memoir manga outstanding and popular. Through this strategy, the author created the "provoker," the other self, who calmly observes the situation from the third-party, bird's-eye view.

Role of Humor

The studies cited in this chapter have brought up the important role of humor in tōjisha narratives. There seems to be a symbiotic relationship between laughing and healing, as some scholars discuss the role of humor in disability memoirs. For example, Yamashita (2009) states that the essence of humor is "a time lag"; that is, humor is able to induce a psychological shift. A quick emotional transposition occurs when we are furious at first and then burst into laughter at someone's "smarty-pants" remark. Without that sort of emotional transition by humor, it is hard to distance ourselves from troublesome situations that we happen to fall into. When a *mangaka* like Komame Daruma creates an autobiographical comic, she resorts to her metacognitive ability to see her experience from a different point of view. Yamashita, a psychiatrist, describes how her own patient, a male schizophrenic tōjisha in his twenties, happily showed his *yon-koma* manga (four-panel manga) story based on his illness experience. The healing power of humor in tōjisha narratives has been pointed out in the studies cited in this chapter. Some of the *tōjisha manga* artists I interviewed told me that creating their manga was their sublimation. As Bingham and Green (2016) mention in the introduction chapter of *Seriously Funny*, disability humor has been a powerful influence throughout history. Here, Japanese Jungian psychologist Kawai Hayao cautions that affection must accompany humor since the act of laughing can easily hurt the targeted person when it comes to a mental problem.

Another important point Yamashita (2009) makes about humor is that to make readers laugh, the author must first make them feel empathy with the author's tōjisha background. Only then can readers share the humor with the author and experience its "mental shift" effect. I believe that this shared humor is what makes *tōjisha manga* special. Yamashita argues that essay manga offers *ikiru chie* (wisdom of life), especially what humor can do for us in dealing with mental disability, and that this genre is indeed a ground-breaking class of comics that specializes in making good use of disability humor. On this topic, I should mention that the author's own portrayal of "useless self" or "miserable self" through a feeble, yet adorable *yuru-kyara* character in this genre makes it even easier to evoke the reader's empathy. Similarly, Natsume (2006) mentions Kera Keiko's gag manga, *My Family* (*Atashin'chi*), and reveals how much he, as a reader, benefited from Kera's manga during the time of his own hardship: The series made him laugh a lot and "saved" him from being totally consumed by emotional pains from caring for his elderly parent (244). Kera's manga, serialized in 1994–2012, centers around a comedic housewife of the Tachibana family. Natsume emphasizes the importance of finding humor in any measures, including reading manga, as laughter often helps us reset our mood and perspective on worrisome affairs (248). I personally attest to the power of humor because I, too, feel that humor can act like *oburāto*, a wafer sheet for bitter medicine (an edible film used to wrap a powdery drug in Japan) that helps the reader to more easily "swallow" a harsh tōjisha narrative.

Hata (2013) further analyzes a critical role that laughter plays in essay manga which narrate horrendous, tragic experiences such as illness—thus, *tōjisha manga*—from both the reader and writer's point of view. The laughter embedded in the storytelling very likely lightens things up, preventing the serious content of the story from becoming overwhelmingly grievous. It also ensures the story to be a "safe" read, where humor serves like a built-in buffer zone between the reader and the tragedy. If a story that has no such buffer tells about, say, the disastrous consequences of eating disorders in an unforgiving way, readers may be easily overwhelmed by negative emotions generated through the manga. Hata also points out that this type of essay manga tends to put more weight on humor over the author's artistic originality or ability to develop an ingenious plot.

From both Yamashita's (2009) and Hata's (2013) studies of autobiographical comics, we see that the emphasis on, and prudent application of, humor is a strength of *tōjisha manga*. Now, Nishikura (2017) elaborates on two types of humor: black humor used purely to entertain and the humor that saves others. In creating a comical disability memoir, Komame used the latter type of humor to save herself. This humor not only rescues herself but also enables other tōjisha to go on with their lives. Nishikura finds that laughing helps us deal with serious challenges in life.

However, we must caution that humor is a double-edged sword. Nishikura (2017) highlights the limited role of humor. Komame came to realize that a gag manga series has little or no room for depicting what truly bothers her as the tōjisha because the audience demands a constant supply of laughter from the story. She compared her work with two non-gag manga titles which seem, in her view, better equipped to discuss the issue at hand more seriously. Because of the nature of this genre, she was obliged to end an episode with some sort of a punchline without ever reaching the point of truth and sometimes felt smothered by that expectation. It was painful for her not to be able to honestly and solemnly convey to her audience the truth that she was really suffering. She admitted that it was her own choice to publish it in the gag manga outlet. But the truth of the matter was that her experience was far from being funny.

Komame sometimes burst into tears when drawing some scenes. Her "laughter" was invented at the sacrifice of her painful, uncontrollable experience. And her audience did not know any of it. Komame was equally bothered by the fact that she was unable to change some readers' laissez-faire attitudes toward the disease. She felt defeated since she had not fully connected with her audiences. Komame's reflection indicates that although powerful, humor is not a panacea for reducing disability-related stigma and prejudice in the public. Nishikura (2017) expressed her view to Komame, saying that the approach implies the commercialism of tōjisha as a commodity, using them like stepping stones. Komame responded that to reach out to the general public and raise awareness about an arcane condition such as alopecia areata, the manga must contain the elements of humor and entertainment. In her view, as long as that strategy does not fall short of solving the core problem in society, humor can be a powerful component of *tōjisha manga*. Komame's graphic memoirs, *A Beginner's Guide—Fight! Alopecia Areata* and *Daily Hair Battle: 24-Hour Close Report of a Wig Life!!*, indeed demonstrate that a comic can effectively convey a tōjisha narrative through a good sense of humor and the universal theme of self-development.

Conclusion

Simply put, the term *essay manga* refers to a comic genre of Japan analogous to autobiographical comics in English-speaking countries. This chapter first focused on the rise of this genre in the 1980s and its development in the 1900s and the millennium in Japan. It then examined the characteristics and strengths of representative autobiographical comics, especially ones that portrayed the authors' lived experience of various psychological or neurological conditions, the subgenre I call "tōjisha manga." Prior to my own analysis of selected *tōjisha manga* in Part II, it was necessary to present as a background what other manga studies specialists and academics discern as the common features and benefits of such autobiographical comics. In this chapter, I provided examples of analysis such as essay manga featuring mental health problems, such as Azuma Hideo's (2005b) *Shissō Nikki* that depicted his lived experience of alcoholism, bipolar disorder, and anxiety disorders, or an illness such as Komame Daruma's *A*

Beginner's Guide—Fight! Alopecia Areata. The important role of humor in tōjisha narratives was also discussed because it is an inseparable topic of *tōjisha manga* as I often describe how humor is embedded in each title.

Nishikura (2017) describes a recent development of essay manga as follows: At first, most essay manga were "*kokuhaku* (confession)" stories, but within the past fifteen years, that trend of the tōjisha narrative changed to the *kyōkan* (empathy) approach. The same shift occurred in the literary world and can be seen in books such as Mizuno Keiya's (2017) *Unbeaten by Deformity*, which inspired and showed the general reader (especially non-tōjisha) hints on how to overcome any difficulty or crisis in life. From stories written in the approach of universal themes such as pain and self-esteem, those who have no tōjisha experience may not necessarily find tōjisha narratives relatable but can still gain empathy. As Nishikura (2017) points out, inclusivity has a positive impact on activism. She reports on a case in which, after a non-tōjisha joined a tōjisha group, the group stopped looking at society as the sheer enemy, and their initial hostility and combativeness may disappear. With that internal change, even other reluctant tōjisha, including those who lacked the passion to reform the society, showed interest in joining the group. This example shows how no boundary (tōjisha vs. non-tōjisha) welcomes everyone. Following this new direction of essay manga, the *tōjisha manga* titles I introduce in Part II attempt to get the majority (non-tōjisha readers) involved by employing the *kyōkan* approach of highlighting our commonality and showing how to solve such universal, relatable human problems as relationship troubles or work and family issues.

Notes

1. This statement appeared on the home page of Media Arts Current Contents, a website that posts essays and interviews about media arts including manga written by recognized experts.
2. It is important to clarify that these scholars examined autobiographical manga published in Japan. *The Four Students Manga*, a Japanese language comic Henry Kiyama self-published in San Francisco in 1931, may be the first autobiographical manga overseas. Similarly, *Barefoot Gen*, Nakazawa Keiji's manga, which was serialized from 1973 to 1987, is probably the best-known autobiographical comic of Japan in the West.
3. *Pixiv Essay* runs a contest called "Pixiv Essay Rookie of the Year Award" with KADOKAWA, which serves as a gateway to manga career success. The contest's winners receive a monetary award and a contract for both online and print publications.

4. Sakura Momoko died from breast cancer in 2019 at the young age of fifty-three.
5. *MOE* (https://www.moe-web.jp/) is a Japanese monthly magazine published by Hakusensha for the fans of graphic novels, including children's picture books, illustrated essays, and manga. In this 2011 article, the name of the writer (or multiple writers) was not disclosed.
6. As *MOE*'s writer mentioned, traveling was one of the most popular essay manga topics at the genre's developmental stage in the late 1980s. Chu and Coffey (2015), the only English-language academic paper I could find that studied this genre, offer two case studies of traveling essay manga using *Hitoritabi ichinensei* (Rookie Lone Traveler) by Japanese illustrator Takagi Naoko in 2006 and *Merhaba! My Turkey Journey* by Taiwanese geography teacher Peiyu Chang in 2009. (The latter title is reminiscent of Takahashi Yukari's 1991 essay comic, *My Turkey Memoir.*) Chu and Coffey claim that "since the mid-2000s, there has been a growing number of autobiographical travel graphic novels depicting independent women artists/travelers" (145). However, if MOE's (2011) analysis is accurate, it seems that female travelers' essay manga actually began in the late 1980s, as represented in Saibara's and Takahashi's works. Regardless, Chu and Coffey's study is promising as the authors attempt some semiotic discussions using such topics as color symbolism and multimodal text analysis via multiple semiotic elements. Especially due to the aforementioned absence of research on essay manga, I hope more researchers will explore essay manga titles using semiotic analysis.
7. Self-narrative was part of postmodernism characteristics developed in the 1960 and 1970s not just in novels but also in movies in Japan. In the field of literature, the very first *shishōsetsu* (I-novels) are believed to be Shimazaki Tōson's *Hakai* (*The Broken Commandment*) in 1906 and Tayama Katai's *Futon* (*Bedquilt*) in 1907. It is also important to point out that the autobiographical form is not unique to comics in Japan.
8. From a gender studies perspective, we should recognize this unique timing and investigate the role that gender played in the development of essay manga. Thus, as I urged before, more comic studies scholars should investigate the genre of essay manga.
9. However, it is important to note that heavy use of background texts might just be characteristic of *My Darling*, not of all essay manga. For example, in *My Darling*, cross-cultural footnotes are needed in the narration to explain the characters' remarks as an interracial couple. In fact, heavy use of handwritten text in some essay manga like Ikeda Kyoko's *What's Up with New Depression?* might tire readers, after wading through lots of text in crowded panels. In contrast, essay manga titles like Fuzuki Fu's *Mama is a Schizophrenia* present only limited handwritten text. Therefore, some of Narioka's findings may not be applicable to the whole genre.

10. I should mention that some manga artists switch their drawing styles according to the genre they choose to present. For example, Oguri's drawing style changed between her *shōjo manga* titled *Kanayako* and her essay manga *My Darling*. In *Kanayako*, Oguri drew characters with more complex features typical of the *shōjo manga* style. Another example is Arakawa Hiromu, a female manga artist known for *Fullmetal Alchemist* (*Harigane no renkin-shi*) and *Silver Spoon* (*Gin no saji*), popular works of *shōnen manga* (comics targeting young male readers). As mentioned in this chapter, for a comic series *Farmer Nobles* (*Hyakushō kizoku*), Arakawa adopted essay manga's humorous simplicity and created amusing stories narrated by a comical, bespectacled bovine character representing herself. Each episode was produced based on Arakawa's experience as a farmhand in her homeland of Hokkaido, illustrating the ins and outs of real farming life with humor, while also revealing serious issues such as the wasteful practice of surplus milk disposal.
11. A clever visualization of the heroine's unspoken feelings and inner voices is demonstrated in fictional *shōjo manga* stories like *The Star of Cottonland* (*Wata on Kunihoshi*) by Oshima Yumiko, one of the twenty-four *nen-gumi* (a group of popular female *mangaka* to whom the *shōjo manga*'s rapid development in the 1980s is credited). Thus, this characteristic was not invented for the *josē* manga genre per se.
12. Regarding anthropomorphism, a study by Sakamoto et al. (2002) offers intriguing findings relevant to the characteristics of this genre although it is a study about essay manga. Using a "ComicDiary," a digital comic diary project, college students were required to report their real-life conference participation as a class assignment. For their diary, the students digitally created a character as a narrator/reporter to represent themselves and conceptualized it as a *jibun no bunshin* (the alter ego), or their own avatar, so to speak. For example, one student drew a "kettle" character (a stubby kettle body with tiny arms and legs) as a self-portrait in his comic diary, in which the kettle man speaks in front of his audience or listens to and comments on someone's talk. Moreover, the students reported that, as they created their comic story, they presented only a digest of their experience to ensure that their report would be approachable and fun to read. In addition, students stated that they purposefully exaggerated their feelings and reactions to make them more noticeable to the reader. In the case of the kettle man, in response to the audience's flattering comments, the character is drawn with a big grin on his face and a full head of steam coming out of his lid, looking like an overheated kettle. Sakamoto et al. found that amplification of emotional expressions was also intended for entertainment: the students desired to make their comic diaries memorable and worth reading by overdramatizing some scenes as *ochi* (punch lines). These

findings have an uncanny resemblance to what was reported by the analysts of essay manga, implying that today's consumers of manga like these college students might have already emulated essay comic strategies and applied them in crafting their own diaries.

13. Although Hosokawa and Nakamura are caregiver tōjisha, Kodaira and Ito include their names in their "manga by tōjisha" list. As mentioned in chapter "Tōjisha", graphic memoirs drawn by family caregiver tōjisha are typically juxtaposed with those of works by *tōjisha honnin* (individuals who are diagnosed with the condition).
14. I would like to mention two additional titles in association with their recommendations. First, Ōhara Yukiko's 2006 essay manga *The Husband of the Ōhara Family* (*Ōhara-san chi no danna san*) is where the aforementioned Ōhara Kōki appears as the title character. In their 2009 paper, Kodaira and Ito comment that this manga uniquely depicts her challenging life married to Kōki who has panic attacks and other neurotic disorders. Due to individual artistic differences, Kōki in Yukiko's manga is rendered differently than Fujiomi's image, even though both figures represent the same person and share common essay manga features such as the plainly drawn character with a disproportionately large head in the comically deformed character style of *santō-shin*. In 2013, Fujiomi Shūko published another essay manga, *My Bipolar* (*Sōutsu nan desu, watashi*), based on her Bipolar II condition.
15. As I will show in the case study chapters, the authors of *tōjisha manga* cleverly utilize their penname as *mangaka* (e.g., Okita Bakka) as well as their avatar character's name (e.g., Bakka-chan) like a double-sided shield which helps reduce the potential harm to the author. The identity of their family members, especially those with lived experience of mental health problems, is more likely protected with the use of pseudonyms in *tōjisha manga*.
16. Insertion of a third-party observer is not uncommon in comics. Chute (2017) explains the technique of "double-trackedness" used in Justin Green's autobiographical comic, *Binky Brown Meets the Holy Virgin Mary.* In this comic, Green, who has OCD, depicts the character Binky's urge to be obsessed with something, while the author's rational voice criticizes how absurd the act is. For example, in the scene where the adolescent "Binky," a young version of the author, is running down the street, the author's retrospective narration also appears above the panel. Green himself called this technique the "double vision" (him-now and him-then), explaining that it enabled him to portray the most private moments from his life. Chute (2017) calls it the double-trackedness, the author's commentary in text boxes paired with the action of the scene upfolding, and aptly explains that this technique creates a tension between the author's retrospective voice and the young self's voice/thought.

References

Arakawa, Hiromu. 2009. *Hyakushō kizoku* (*Farming Nobles*). Volume 1. Tokyo: Shinshokan.

Asano, Tomohiko. 2001. *Proximity to Self through Narratives—From Family Therapy to Sociology* (*Jiko e no monogatariron-teki sekkin—kazokuryōhō kara shakaigaku e*). Tokyo: Keisō Shobō.

Azuma, Hideo. 2005a. "'I resurrected by writing a true-life story.' The Legendary Comic Artist Azuma Hideo (*Jitsuwa wo kaite fukkatsu shimashita. Densetsu no mangaka Azuma Hideo*)." *Shukan Asahi*. July 1, 2005: 34–36.

———. 2005b. *Shissō Nikki*. Tokyo: East Press.

———. 2006. *Utsu Utsu Hideo Nikki* (*Melancholy Hideo Diary*). Tokyo: Kadokawa Shoten.

Bingham, Shawn Chandler and Sara E. Green. 2016. *Seriously Funny: Paradoxical Power of Humor*. Lynne Rienner Publishers: Boulder, CO.

Chu, Kin-wai, and Simon Coffey. 2015. "Multimodal Analysis of Graphic Novels: A Case Study Featuring Two Asian Women Travelers," *Intercultural Communication Studies* XXIV (1): 145–165.

Chute, Hillary. 2017. *Why Comics?: From Underground to Everywhere*. HarperCollins: New York, NY.

Fujiko, Fujio. 1978. *Pāman no hibi* (*Daily life of Parman*). Tokyo: Shogakukan.

Hanawa, Kazuichi. 2018. *Keimusho no naka* (*Incarcerated*). Tokyo: Kodansha.

Hasegawa, Machiko. 2016. *Sazaesan Uchiakebanashi* (*A Memoir of Drawing* Sazaesan). Tokyo: Asahi Shinbun Shuppan.

Hata, Mikako. 2013. "Drawing 'Depression'—Examination of the potentials of comic representation." *Joshigaku Kenkyu/Konan Women's College Women's Studies Research Club 2*: March, pp. 66–75.

Isayama, Motoko. 2018. *Atai bijinbyōni narimashita* (How I got an illness that made me beautiful). Tokyo: Kodansha.

Kawahama, Hideaki. 2007. "Essay Manga's Process of Establishment—Representations of Manga Artists Themselves in Special Issues of Shōjo Comics (*Essei manga no seiritsu katei—bessatsu shōjo comics ni okeru mangaka no hyōsō ni chakumokushite*)." *Manga Kenkyū*, No.12, 62–70.

Kodaira, Tomoe, and Takehiko Ito. 2008. "Mental Disability Memoirs—Multiple Meanings of Narratives (*Seishin shōgai no tōbyōki—tayōna monogatari no igi*)." *Macro Counseling Kenkyū* No.7, 48–63.

Komame, Daruma. 2012. *A Beginner's Guide—Fight! Alopecia Areata* (*Watashi, pika pika no ichinensei—Dato! Enkei datsumōshō*). Tokyo: Kadokawa Shoten.

———. 2016. *Daily Hair Battle: 24-Hour Close Report of a Wig Life!!* (*Hibi zurezure: Katsura seikatsu micchaku 24-ji!!*). Tokyo: Shōgakukan.

———. 2013. *My Husband is a Burmese: Secrets of the World's Happiest Country* (*Danna sama wa būtan jin: Sekai-ichi kōfukuna kuni no himitsu*). Tokyo: KADOKAWA.

Kondo, Masataka. 2006. "Saibara Rieko and manga journalism." In July issue, "Saibara Rieko Special," *Yuirika*. Tokyo: Seidosha.

Kubota, Yasutaka. 2006. "Mental health related topics seen in today's manga culture (*Konnichi no manga bunka nimiru mentaru herusu kanren no wadai*)." *Shiga Health Report* No. 63: 3–4. Shiga University Health Maintenance Center.

Kurata, Mayumi. 2001. *Loser-Men's Walker* (*Damenzu-Wuōkā*). Tokyo: Fushosha.

Kuruneko, Yamato. 2006. *Kuru Neko*. Volume 1. Tokyo: KADOKAWA.

Matsui, Takeshi. 2011. "Analyzing Consumer Culture Dynamics through Text Analysis of Media Discourse: A Sociological Approach." In *European Advances in Consumer Research*, 9, edited by Alan Bradshaw, Chris Hackley, and Pauline Maclaran, pp. 562–563. Duluth MN: Association for Consumer Research.

Mizuno, Keiya. 2017. *Unbeaten by Deformity—Nine Stories of Achieving Happiness in Spite of Unusual Appearances* (*Kao nimo makezu—Donna 'mitame' demo shiawase ni nareru koto wo shōmeishita kyūnin no monogatari*. Tokyo: Mizuno Office.

MOE. 2011. "Comic Essay's Great Power (Komikku essei no sugoi jitsuryoku)." Monthly MOE 33(1) No. 375, January 2011: 12–41. Tokyo: Hakusensha. (no specific writer's name noted).

Nakamura, Yuki. 2008. *Wagaya no haha wa byōkidesu* (My Mother is a Sicko), Tokyo: Sanmāku Shuppan.

Narioka, Keiko. 2012. "Essay Manga's Multimodality: Visualization of Textual Information and 'Between-Genre'" (*Essei manga no maruchimodaritei: Gengo jōhō no bijuaruka to 'kan-janrusei'*) *Toyo University Repository for Academic Resources* [Toyo Hōgaku], 56(1): 358–333.

Natsume, Fusanosuke. 2006. *What's Wrong with Learning about Life from Manga?* (*Manga ni jinsei wo manande nani ga warui?*). Tokyo: Random House Kodansha.

Nishikura, Miki. 2017. "Humor for Survival—Laughing at Oneself, Lightening Everyday Struggle. (*Ikiru tameno yūmoa—Jiko wo warau, yamai no nichijo wo warau*)." *The Horizon of Storytelling: Life Story Review. Journal of Japan Live Story Research Institute*, 2: 3–24.

Oguri, Saori. 2002. *My Darling is a Foreigner* (*Dārin wa Gaikokujin*). Tokyo: KADOKAWA.

Ōhara, Yukiko. 2009. *Mr. Ōhara and His Recent Neurosis* (*Ōhara-san chi no danna sann konogoro sukoshi shinkeishō*). Tokyo: Bungei Shunju.

Ōhara, Kōki, and Shūko Fujiomi. 2002. *Seishinka ni iko!* (*Let's Go to a Mental Health Clinic!*), Volume 1. Tokyo: Bungei Shunju.

Oshima, Yumi. 1978. *My Toy Box* (*Yumin omochabako*). Tokyo: Puchikomikku.

Pixiv Essay. 2016. Why is Essay Manga Popular in the Social Media? New Business Tactics of Publishers (*Essei manga ga SNS de ninki no riyu towa? Shuppangyokai mo aratana ugoki* 2016). June, 13. https://ddnavi.com/news/305909/a/.

Sakamoto, Ryuuki, Yasuyuki Sumi, Keiko Nakao, Kenji Mase, and Susumu Kunifuji. 2002. "*ComicDiary: Supporting Exchange of Individual Experiences and Interests in Comics Style* (*Kommikudaiari: Manga hyōgen wo riyōshita keiken ya kyōmi no dentatsu shien*)." *JAIST* (Japan Advanced Institute of Science and Technology) *Repository*, 43(12): 3532–3595.

Saida, Takahiko. 2001. "Shishosetsu to watakushi manga." *Sekai no Bungaku* No. 110 (pages unknown).

Saibara, Rieko. 1990. *A Wanderer's Diary of Mahjong* (*Mājan hōrō-ki 1*). Volume 1. Tokyo: Takeshobo.

———. 2004. *Mainichi Kāsan* (*Being a Mom Every day*). Tokyo: Mainichi Shinbun Shuppan.

Sakura, Momoko. 1987. *Chibi Maruko-chan* Volume 1. Tokyo: Shueisha.

Sato, Miyuki. 2013. *Nanbyō nikki* (*A Diary about an Intractable Disease*). Tokyo: Taiyo Toshokan.

Takahashi, Yukari. 1992. *My Turkey Memoir* (*Toruko de watashi mo kangaeta*). Shueisha: Tokyo, Japan.

Takemiya, Keiko; Uchida, Tatsuru. 2014. *Take and Tatsuru's Manga Culture Theory* (*Take to Tatsuru no manga bunka-ron*). Tokyo: Chūō Seihan Inc.

Takeuchi, Miho. 2019. "Essay Manga Genre and Manga Artist Sakura Momoko—From 'Manga Version, *One-Man Mimicry of Wrestling*' (*Essei manga toiu janru to Sakura Momoko—'manga-ban hitorizumo' kara*)." *Medial Arts Current Contents*. https://mediag.bunka.go.jp/article/article-14951/

Tan, Caroline S.L. 2014. "*Gotochikyara* & *Yurukyara:* The Fusion of Pop Culture in Place Branding in Japan." *International Journal of Business and Economics*, 6(1): 75–89.

Tomoyo. 2018. *Watashi chūgakusei kara tōgōshicchō yattemasu*. Tokyo: Gōdō Shuppan.

Yamagishi, Ryoko. 1973. *Ghost Tales—My Ghost Experience* (*Yureidan—Jibun no reitaiken*). Tokyo: Shogakukan.

Yamashita, Keiko. 2009. "Reading Comic Essay (*Komikku essay wo yomu*)." *Japanese Journal of Clinical Psychology*, 9(5): 698–701.

Yoshitani. 2007. *I, Otaku Salaryman* (*Boku, Otariiman*). Tokyo: Chukeishuppan.

PART II

Case Studies

CHAPTER 5

Okita Bakka's *Gaki-tame* Series (2011–2013): A Memoir of a "Troublemaker" Aspie Girl

Introduction

This chapter showcases Okita Bakka's *tōjisha* manga, *Gaki no tameiki*, through Kiss, Kodansha's division of a monthly magazine for *josei* manga (manga for manure female readers) (Fig. 5.1). A three-volume series published in a span of three years (volume 1 in 2011, volume 2 in 2012, and volume 3 in 2013), the story features Okita's lived experience with *hattatsu shōgai*. This is a visually innovative graphic autobiography about Okita's days as an elementary schoolgirl who was constantly labeled as a "*mondai-ji* (problem child)" by teachers. The forty-year-old manga artist Okita met with me for the interview while accompanied by her *tantō henshūsha* (manga editor), Ito Norikazu. Since Okita and Ito often used *Gaki-tame*, the shorter version of *Gaki no tameiki*, I will use *Gaki-tame* hereafter.

Hattasu shōgai is a three-pronged neurological condition that entails Asperger's syndrome (AS), learning disability (LD), and attention deficit hyperactivity disorder (ADHD). Although Okita was first diagnosed with LD and ADHD in the fourth grade, it was not until adulthood that her Asperger's syndrome was officially identified. The Japanese term *hattatsu shōgai* can be translated into English as "developmental disabilities." However, it is not the same as neurodevelopmental disabilities defined in *DSM-5*, which include intellectual disability and language disorder (2013,

Y. Okuyama, *Tōjisha Manga*,
https://doi.org/10.1007/978-3-031-00840-5_5

Fig. 5.1 Image by Okita Bakka. (Courtesy of Kodansha)

p. 31). Kayama and Haight (2014) describe *hattatsu shōgai* as "relatively mild cognitive and behavioral disabilities" and argue that to understand this condition "requires examination of Japanese social, historical, and cultural context" (p. 3). Children who are diagnosed with this condition typically have relatively mild impairments of cognition, attention, and executive functioning. In the context of Japanese schools, these children's behaviors are typically viewed as lazy, uncooperative, or selfish. To clarify, *hattasu shōgai*, which includes Asperger's syndrome, is not a mental illness but a neurological condition.

During the interview, Okita referred to her medical diagnosis as AS (pronounced "eiesu" in Japanese) most of the time. Therefore, mostly AS and *hattatsu shōgai*, and occasionally ASD (Autism Spectrum Disorder), are used in this chapter.[1] In the interview, Okita provided deep insight into how *hattasu shōgai* has affected her self-esteem. For those who are diagnosed with ASD, three subtypes have been identified: *fuchūi yūseigata* (carelessness-dominant model), *tadōsei/shōdosei yūseigata* (hyperactivity/impulsiveness-dominant model), and *kongōgata* (mixture model). Based on her self-identified characteristics given in the interview, she appears to belong to the last type. When she was diagnosed with LD and ADHD in the fourth grade, the term, *gakushū shōgai* (learning disability), had left nothing but a negative impression on her. "I thought that meant I'm stupid," Okita told me. To the child Okita, it sounded like a permanent stamp for "*benkyō ga dekinai*" (poor at schoolwork) children. Detailed descriptions of her *hattatsu shōgai* traits can be found in the sections, "The Author's Background" and "Drawing Styles and Storytelling Characteristics."

Gaki-tame is, according to the editor Ito, "a long-run (*iki no nagai*) manga," the type of manga estimated to keep getting reprinted for many years. While writing *Gaki-tame*, Okita was simultaneously working on two more graphic memoirs, a single-volume manga *Nitoro-chan* (Little Nitoro, 2013a) published by Kobunsha in 2013 and a seven-volume series *Mainichi yarakashitemasu* (*Everyday life of an adult Aspie*, 2013b), or *Yarakashi* for short, by Bunkasha. Still serializing since 2013, *Yarakashi* illustrates how Okita's adulthood is affected by the condition of *hattatsu shōgai*, and the other two focus on her childhood experiences. "Those who know me think I'm 'a *mechakuchana* (messy/incoherent) person,' even though I think I'm a serious person," she said laughingly. On a serious note, she emphasized "a perceptual gap" between the *kenjōsha* and the *tōjisha*. With *Yarakashi*, "I wanted to depict these incongruous views of me. I can't behave like others, and they can't behave like me." Unlike *Gaki-tame*, *Yarakashi* is the manga that focuses on what she, an adult Aspie, experiences, "including my record of wrecking five cars in a year because I fell asleep at the wheel (*inemuri unten*)," she said, laughing.

Okita calls *Nitoro-chan* "the first manga about AS written from the *tōjisha*'s point of view," and considers it "a dark story" because up until then, she had published stories that were replete with jokes and slapstick actions in the genre of gag manga. However, she received a fairly good audience reception, and *Nitoro-chan* turned out to be the first of her three *hattatsu shōgai* memoirs that she believes "helped raised public awareness and popularized the term *hattatsu shōgai*." She stressed that people's familiarity with the term gradually increased because of that book. However, *Nitoro-chan is* a comic with no sequels, not containing as much content and artistically more developed drawings as *Gaki-tame*.[2] Therefore, I chose the latter to analyze as the *tōjisha manga* on childhood AS.

The Title and Target Readership

The original Japanese title of *Gaki no tameiki* is ガキのためいき (*A Kid's Sighs*). The title *Gaki no tame iki* was entirely Okita's idea. Ito confirmed, saying, "It [her proposed title] stood out because little kids don't usually give out a sigh. My impression was that it's quite an interesting title!" The title has two key words: *gaki* (kid, brat) and *tameiki* (sigh). In the interview, Okita first explained the meaning of "sigh" (*tameiki*)[3] in the title. The protagonist, Bakka-chan, is the childhood Okita, who does not act like an ordinary child. "Yes, kids normally don't sigh," Okita said, but Bakka-chan does sigh in numerous scenes when she is distressed by her inability to explain in her own words to teachers and classmates, even to her mother, why she acts the way she does. Furthermore, Okita wanted a "*katakurushiku nai*" (unstilted/informal) title for her graphic memoir. For her choice of the word *gaki* (kid, brat), she explained that it is a story about an "atypical" child and that, as an example, Bakka-chan exerts a bizarre physical reaction to other young children's high-pitched voices (to which she is actually hypersensitive). "From her teachers' viewpoint, she is more like a *gaki* (brat) than a *kodomo* (child). In fact, they consider her a *kuso-gaki* (crap brat). So, I thought the word *gaki* is more fitting," she added. In essence, Okita wanted to tell a story about a girl labeled "a troublemaking child (*mondai-ji*)" in school, who is, in fact, experiencing sorrows and distress over the situation herself. She said, "When thinking of that theme, I found myself sighing instantly." A little child is not equipped with the ability to account for what is going on in her mind and how autism is affecting her brain; yet, she is called unfairly *baka* (a moron), *kimoi* (weirdo), and *kirai* (despicable) by everyone around her. Recalling all these memories, she could do nothing but sigh. "It felt like the child is perceived as incompetent, except for breathing. The meaning of *tameiki* (sigh) in the title has those kinds of connotations."

Okita's insight raises the issue that those who had no knowledge of *hattatsu shōgai* back then viewed her as someone sucking air uselessly. She also told me that she was an absentminded child and spent her day imagining things alone. Not knowing about her ADHD-affected concentration, her teachers and other adults constantly scolded her, which led her to believe she was "*dame na kodomo*" (a child of no good). Okita eventually developed *bamen kanmokushō* (selective mutism) in the third grade, which further aggravated her inability to defend herself from the label of a troublemaker. "My childhood was basically void of anything enjoyable. That sentiment is reflected in the title," she told me, stressing, "The word *tameiki* summarizes it all."

The target readership of *Gaki-tame* appears to be both *hattatsu shōgai* tōjisha and their family, although Okita told me that she had no particular readership in mind when creating this work. This manga illustrates how she had struggled with the *shūdan seikatsu* (group-oriented culture) of the Japanese school system ever since she entered elementary school. Okita learned from readers' comments that teachers are still not properly trained to understand what causes the behaviors that clash with the school's *shūdan seikatsu* in children with *hattatsu shōgai*; therefore, she is particularly happy to hear from many schoolteachers who read *Gaki-tame* and her other two graphic memoirs. To this day, teachers are struggling to figure out how to work with children like Bakka-chan who rock their chairs wildly or keep moving around in the classroom. These comments inform Okita that when teachers read the *Gaki-tame* series, especially Bakka-chan's thoughts, they begin to understand what Autism Spectrum Disorder, or ASD, is.

Another group of readers who found *Gaki-tame* helpful are children with *hattatsu shōgai* themselves. For example, Okita heard from one of her former classmates who had a child diagnosed with *hattatsu shōgai*. When adults told him in the third grade, 'You have *hattatsu shōgai*,' he had a hard time understanding what it meant. However, when the child read one of Okita's memoirs, he immediately said, "Oh, it's me; it's about me!"

His mother was so touched by that that she cried and thanked Okita profusely. As with this case, it is fairly common for individuals to learn about their undiagnosed ASD by first reading a memoir by a *hattatsu shōgai* tōjisha in Japan (Tateiwa 2014). Okita said to me, "This is perhaps an example where my manga helps *tōjisha* see objectively how they get in trouble because of their *hattatsu shōgai* traits." However, Okita admits that *Gaki-tame* depicts the experience of an Aspie schoolgirl in the 1970s and that since then, the knowledge of *hattatsu shōgai*, including *ryōiku* (special training for children with *hattatsu shōgai*),[4] has progressed rapidly. Fortunately, after her pioneering memoirs on childhood autism were published in the early 2010s, more manga about *hattatsu shōgai*, either autobiographical or fictional, have been written by other manga artists.

The Author's Background

Her penname, Okita Bakka (沖田X華), is derived from the Japanese phrase, *okita bakkari* ("just woke up"). This "half-asleep, half-awake" image of Okita is represented as a *bōtoshita ko* (an absentminded child) in *Gaki-tame* and as an adult who has difficulty paying attention to tasks and loses personal items such as keys and credit cards (particularly during her menstrual period) in *Yarakashi*. The letter X of her penname connotes a "wrong," "failure," or "damage" in Japanese culture. For example, incorrect answers on a test are marked with an X (e.g., a phrase, *batsu ichi*, or "damaged once," is a euphemism for "being divorced once" in Japanese). It may be possible that Okita also turned the slander of "*Bakka*!"—*baka* ("a moron") said emphatically—thrown at her by peers and teachers into a memorable *nom de plume* for herself. In the interview, Okita described how children with *hattatsu shōgai* typically have a *mainasu no imēji* (negative self-image) in school. Constantly scolded by adults, "I felt that I was a girl who was very different from others," she said. As depicted in *Nitoro-chan* and *Gaki-tame*, Okita was the only child among many classmates who was unable to engage in *shūdan kōdō* (group activities). She said, "I was pretty convinced that I was unique. [Laugh] I mean, nobody else was labeled *mondai ji* (problem child). Actually, many boys acted similarly. But for girls, I was the only one. It was rare for a girl to be like that." Then, she said jokingly, "I even wondered, 'Maybe I'm a boy,' because I didn't know back then what caused me to behave like that. I even thought that sooner or later, a penis will grow on me."

Okita is originally from Toyama Prefecture. Born in February 1979, she had just turned 40 years old when I interviewed her. Prior to making her debut as a manga artist in 2008, Okita worked in a hospital as a nurse and then in the industry of adult sex entertainment, or *fūzoku* (short for *sei fūzoku*, or literally "sex customs"). Her father left her family after Okita graduated from high school. She has two younger brothers whom she talked about briefly during the interview. Her brother, who is two years younger, is "probably the only neurotypical member of my family," she said. Her second brother was born when she was still in elementary school. She added, "The moment he was born, I thought, 'I don't like him.' [Laugh] It was as if both of us were sensing the same [neurological] condition in each other. [Pause] We didn't grow close." She said that because of his severe condition of *hattatsu shōgai*, he ended up in a group home. Because she and this brother, who is five years younger, did not get along when they were little, they chose to stay out of each other's lives until recently. As adults, she and her brother figured out how to communicate without getting upset at each other's behavior stemming from *hattatsu shōgai*. She also commented that her mom has mild *hattatsu shōgai* like hers. That is an important observation, as this condition is said to have a high occurrence among family members (Iwanami 2017: 29).[5]

In the prefectures with only a limited provision of *ryōiku*, children with LD are often labeled "no-good children" and left untreated. Okita said that unless there is a unilateral quality control within the whole nation, it is hard to resolve the issue of inequitable opportunities of *ryōiku* for children living in different prefectures. Her brother who has *hattatsu shōgai* fell into that crack and became a *hikikomori* (a shut-in).[6] His symptoms were more severe than Okita's, and nobody at school wanted to deal with him. Unlike Okita, he withdrew from interaction with peers and stopped going to school. He currently lives in a group home, and without a high school diploma, he continues to have trouble finding a way to support himself.[7]

Okita met her current husband, Sakuraichi Bāgen, around the time she was working in the *fūzoku* industry and moved to Tokyo to live with him. Sakuraichi (former penname, Sakurai Toshifumi) is also a manga artist with a longer career although their artistic genres are different. Okita and Sakuraichi work in their own rooms at home and rarely comment on each other's work. However, once in a while, Okita asks him, "Does this drawing make sense to you?" Okita told me that she asks for Sakuraichi's advice only if her *nēmu* (storyboarding) is not making progress. Initially, Okita wrote *ecchina* (erotic) gag manga, a genre called *ero-gyagu* (erotic gag manga) specifically for erotic manga stories, the genre Sakuraichi specializes in. At that time, her manga was mostly a *yon-koma* (four-panel) manga. "I think writing about people is my cup of tea. I'm really interested in other people, in knowing about their past," she told me. As for her three serious yet humorous memoirs of *hattatsu shōgai*, she did not intentionally make them funny, but because her manga career began in gag manga, humor is always part of her storytelling style. Like all manga artists, Okita has received criticism from readers, too. However, she asserted that her desire to inform the public about *hattatsu shōgai* exceeds her fear of losing privacy.

As mentioned before, Okita once worked part-time in the sex industry. That was when she began to consciously explore the meaning of storytelling for entertainment. Furthermore, she initially thought the easiest way to create manga was to employ the *yon-koma* manga format. "Especially because I was not good at drawing, I thought I could get away with my limited [drawing] skills in *yon-koma* manga," she told me. However, that assumption turned out to be false; she realized then that she was "a *mangaka* who knew nothing about manga." Fortunately, Okita was gifted with something crucial to manga artists: according to her husband, she possesses a natural talent for storytelling. That talent is "something you are born with," she told me with confidence. To her, talent is a sense with which one knows what good storytelling is; thus, a gifted *mangaka* feels it in their bones. She stressed that to make a story worthy of reading, it is insufficient to just write about "*hontō no koto*" (what actually happened). From a decade of experience working in this industry, she said she had also learned that "stories don't sell if you just write what actually happened. Manga is actually more than that."

I asked Okita why manga allows her to express things the way she does. After contemplating, Okita said, "I once wrote something only with words. But I was told, 'You really have no literary talent' (*hontō ni bunsai nai*). [Laugh] A simple answer to your question would be that this [manga] is the medium that fits me most (*ichiban jibun ni atteta*), I think." Furthermore, she said that drawing had been her favorite pastime since childhood as depicted in *Gaki-tame* where Bakka-chan happily draws pictures in isolation. However, she avoided drawing figures below the face back then. Laughing, she said, "I can draw the same face over and over, but somehow, it's hard to draw beyond that. I'm terrible with fingers—If I draw them, the image often reverses [the thumb drawn in the place of a pinky, for instance]. So, I usually end up having my assistants do that work." Okita told me, "Somehow, I always find it easier to read *seinen-shi* [manga magazines for young adult male readers]. I like the kind of manga that men in their 60s read. Those are easier for me to understand." In comparison, she finds *shōjo* manga, especially its rom-com stories, very hard to understand. "The characters eyes are *kyon-kyon shitarisuru* ["flamboyant" or "flashy" although I am not quite sure what she means]. They are HUGE. There aren't people like that. At least, not in my hometown," Okita joked. "I just can't understand it. And there are times I'm clueless about how to read it [rom-com manga]. I mean, I'm confused about where to begin reading." As described in chapter "Essay Manga", *shōjo* manga do have more elaborate and complicated layout structures so it takes some reading experience to figure out which frame is the first one to read and so on. However, lack of familiarity is not Okita's issue with the genre as she used to be an avid manga reader. She claims that she cannot get "emotionally invested in the characters."

Ironically, however, it was in the *shōjo* manga division in which Okita received the 42nd Kodansha Manga Award in 2018 for her nine-volume series, *Tomei na yurikago: Sanfujinka-iin kangoshi minarai nikki* (*An Invisible Cradle*: *A diary of a nurse-trainee at an OB-GYN clinic*, 2015), which narrates her part-time job experience at an OB-GYN clinic. She chose to write about it because she learned a lot about abortion from that job and was also told that no *mangaka* had written about abortion, at least in detail. To her surprise, the *Tomei na yurikago* series elicited a huge volume of comments, especially from female readers because pregnancy and giving birth are topics germane to women. The story was later adapted into an NHK (*Nippon Hoso Kyokai*) drama. The TV adaptation by NHK was aired in ten episodes in 2018. When informed of the award, Okita thought to herself, "That's impossible. [Laugh] Especially in the Girls Comics Division!"

Okita is a *hattasu shōgai* tōjisha. As part of her Asperger's syndrome traits, she tends to have vivid memories not only of what happened years ago but also of how she felt in that particular spring in the first grade, for instance. She remembers what got her in trouble more clearly than what made her happy. According to her, "Those memories have been piled up like file folders" in her memory bank. "These are the memories about how and when I got in trouble and how things turned out. I just turned them into manga stories," she told me, smiling. Clearly, her exceptional ability to recall past events, many of which are serious, dark memories including her failed suicide attempt, works to her advantage in her career. I asked her how she makes sure that humor does not disappear totally and darkness does not set in when telling such grim stories to readers; and what sort of trick or device she uses to accomplish that. She answered that she does not consciously think about how bleak her life had been back then. "I'm used to being unappreciated since my childhood anyway. There is this child '*watashi*' (me) character in my head. But my grownup self doesn't get sentimental in revisiting that child character," she said rather triumphantly. In telling her childhood stories, Okita simply retrieves those files from her memory bank without much emotion attached.

It appears that she has the ability to stay emotionally detached from her miserable past. In fact, people often tell her that she has an objective approach toward seeing and depicting herself. Her emotional detachment seems to help her avoid feeling the same magnitude of pain felt by the protagonist, Bakka-chan. She added that psychologically she is "good at bouncing back from a setback, even a serious one that makes one want to commit suicide," referring to an occurrence when she was a nurse. In describing her failed suicide episode, which she wrote about many times in her manga, she said she was rather concerned how *kakkō warui* (awkward) it would have been to survive her suicide attempt at that time. "It's like, 'I screwed up,'" Okita said, laughing. She thought, "Eek, I didn't die!" and, after debating what to do next, she just went to work on the same day. "I'm not sure why, but I cleaned up the room. I asked myself, 'Since I tried and didn't die, does it mean I don't need to die anymore?' [Laugh] Later, I made this incident into a manga episode, and the story didn't sound too grim to my surprise. 'It doesn't appear somber,' people told me." She continued rather calmly, "To me, it's '*mō owatte shimatta koto* (the thing in the past),' so to speak. I'm writing about what happened." Even with some emotional disconnection, her ability to recall in detail haunts her sometimes. However, that emotional reaction ceased after she reframed her memory as "*Mō ikka* (it's over)," she said laughingly. It appears that Okita has developed a healthy distance from her experiences of childhood.

During our interview, Okita shared with me how she found out about her neurological condition. She was in middle school when she was first told that she had Asperger's syndrome. However, it was not until her adulthood that she finally accepted the diagnosis herself. It was because, at that time, "nobody had yet clearly defined Asperger's syndrome, especially its specific symptoms." She added that even today, some of those who received an official diagnosis through trial and error are still seeking to understand what Asperger's really is. "I took my brother's *futōkō* (truancy; refusing to go to school) as a clear example that my case was different (from his and other Aspies)," she said. Because Okita did not develop *futōkō* like her brother, she did not believe that she too had Asperger's syndrome. "Sure, I had trouble following directions and did things in my way (*jibun katte*, or "self-centered"), but I incorrectly attributed it to my blood type, which is B. That's why when I was in school, I was never aware that I had AS [Asperger's syndrome]," she added. However, when she began working after graduating from high school, her AS-specific tendencies (*Asuperugaa no tokusei*) "exploded." That was when she finally came to realize "something wrong was going on (*nanka okashii na*)." She was twenty-three years old then. Prior to that revelation, she had worked as a nurse in a hospital for a few years, during which she could not comprehend what others were telling her to do. Most critically, she made some "highly unlikely (medical) mistakes" (*ariennai shippai*), which eventually made her decide to quit. When she realized she could not process information in ordinary Japanese, she felt something was wrong with her but could not substantiate what was causing that communication trouble.

However, through an online community called Mixi,[8] she eventually learned that all of her work-related trouble was related to ASD. While she was searching for information about ADHD (in Mixi), she ran into the term *Asperger's syndrome*. "Through that venue, I found a site that listed 100 tendencies associated with it. To my surprise, all 100 traits were applicable to me. That's when I first recognized Asperger's as a term, the exact label my mother referred to a while ago." One of the listed traits was *chōkaku kabin* (auditory hypersensitivity). She immediately knew it accounted for her severe reactions to certain sounds. She told me, "When I read about it, everything about it was applicable to me. A bulb went off in my head! I learned that *hattatsu shōgai* doesn't refer to this and that type of person but rather to a neurological condition in which a host of traits you have interact with each other and hamper your daily life." That discovery was made when Okita was about 25 years old. Up until that point, "I had been in the grip of something inexplicable," she added. Thus, she was actually "relieved because the term explained everything." She wanted to meet with other tōjisha, wondering what other adult Aspies were like. Since then, she has attended several "*ofu-kai*" in Tokyo.[9] Through the *ofu-kai*, she got to know people with various types of *hattatsu shōgai* and felt relieved especially when she met other women. What benefited Okita most from these *ofu-kai* meetings was the knowledge that no two people with *hattatsu shōgai* are alike. Even with the same diagnosis, "people grow to be very different adults depending on the circumstances they were born into," she emphasized. As an adult diagnosed with *hattatsu shōgai*, Okita emphasizes the importance of "the division of labor." She said, "There are many things I cannot do. So I pay others to do those tasks for me. Especially, I can't write *kanji* characters correctly, despite the fact that I've practiced ten thousand times (*nanmankai renshushitemo*)." Her assistants also draw landscapes and detailed body parts, things that she is not good at. She believes that rather than laboring over something difficult, "it's much easier to delegate the work to others." With laughter, she said, "It's been very helpful." Many children with *hattatsu shōgai* are known to have trouble learning *kanji* in school.

She identified approximately fifteen Asperger's syndrome traits listed on that site as hers. One of them was the inability to remember people's faces. She learned that because she suffers from *tanki kioku shōgai* (short-term memory disability), she has trouble remembering faces. Strangely, however, that same condition enables one to recall past events clearly and for a long time even though they cannot recite a cell phone number upon hearing it. Laughingly, she said, "I can't remember what I was just told. People tell me to take notes. But I end up losing the notes because of my ADHD. I've lost count of how many schedule books I've lost," she said. People usually struggle because of those specific disabilities, as well as *niji-shōgai* (secondary disabilities) such as depression, developed as a result of *hattatsu shōgai* traits. Another case is that of *tekiō shōgai* (adjustment disability), which prevents some people from getting a job, and they finally get diagnosed in adulthood. Okita stressed that people with *hattatsu shōgai* have different traits and backgrounds. "That particular knowledge that I gained from Mixi was very beneficial to me," Okita said.

There was no particular incident that triggered her coming out as a *hattatsu shōgai tōjisha*. However, besides getting to know other people with *hattatsu shōgai* through Mixi, the publication of Nonami Tsuna's *Akira-san wa asuperugā* appeared to be an incentive for Okita's decision to write *Gaki-tame*, since she said, "By then, someone published a manga about *hattatsu shōgai* called '*Akira-san ga [*wa] asuperugā*,' which was about the person who discovered her partner's Asperger's syndrome after they got married." That manga was created by Nonami Tsuna as an eight-volume series (2011–2017). Okita added that Nonami's series sold fairly well. "When I read it, I came to understand why." Unlike Okita, Akira,

Nonami's husband in *Akira-san* series (covered in chapter "Nonami Tsuna's *Akira-san* Series (2011–2017): A Memoir of a "Cassandra" Wife"), does not have an LD, was an academically strong student, and managed to stay out of trouble in school. But as soon as children were born, his marriage began to go down a rough road. Okita pointed out that the differences between Akira and herself are not only their Asperger's classifications—Akira's is depicted as *judōsei-gata* (passive type) in Nonami's manga—but also gender-related issues.

She said, "It was the first time that I realized that there are differences in men and women with AS. I knew that some female Aspies have published memoirs. But I'm not good at reading books." That is when she decided to share her own experience with the public in the form of manga. As mentioned earlier, she first published *Nitoro-chan*. But in *Gaki-tame*, she focused on the representation of a female case in contrast to Nonami's series, which featured a male Aspie. Okita had searched but could not find any female counterpart in manga and decided to write her own. After the publication of the *Gaki-tame* series, people began to say, "Oh, Okita Bakka, that person with *hattatsu shōgai*," even though she had already come out as an Aspie. "People didn't take me seriously for a while. For example, if I told others that such and such things happened to me, they would say, 'Me, too.' When I was a nurse, especially, my coworkers dismissed my disabilities with 'Nay, nothing special.'" Intuitively, however, she knew that her situation was different. Unfortunately, that difference was invisible to the neurotypical. Therefore, she felt the urge to write about it "to get my point across to these naysayers." She went on, "Their assumptions were wrong, and they didn't listen even when I tried to explain. That frustration prompted me to write this manga and emphasize my viewpoint."

Okita's Drawing Styles and Storytelling Characteristics

Manga is a unique medium that requires a set of layout rules to follow. Normally, manga artists follow rules such as keeping all drawings, including figures and landscapes, within a given panel. However, Okita is a *mangaka* who sticks with a simplistic layout (e.g., see Fig. 5.2), a style quite suitable for the genre of essay manga that I described in chapter "Essay Manga". "People say my manga is one of those that are somewhat easy to read," she said. She added, "I'm kind of scared of going beyond the rules.

3

Fig. 5.2 Image by Okita Bakka. (Courtesy of Kodansha)

I'm not sure why, but it's scary. So, I limit myself to dividing the page into three columns and then each column to halves." For this reason, she does not use complicated layout styles, nor does she jampack images into one panel. Although she claims that she intentionally made *Gaki-tame* comprehensible so that it could "serve as an educational material," her fear of leaving structured patterns when drawing manga might be due to her Asperger's syndrome, if we infer from many episodes of *Gami tame*, which depict Bakka-chan panicking when things go outside of routine (e.g., a sudden change of a schedule). Perhaps she does not intentionally make her manga simplistic or eschew the *gekiga* style (realistic manga drawing style); rather, simplicity may be precisely the outcome of avoiding sudden change and transitions as well as self-regulating emotions through which she has reached a high degree of reader-friendliness. Naturally, all manga Okita has produced have a plain character design (*egara*) and page layout (*komawari*). I asked Okita if children with ASD would find Okita's work easier to comprehend than other manga. Okita cheerfully replied, "I can't draw any better than this, actually," at which point, Ito burst into laughter.

However, there was one change she consciously made in *Gami tame* from her previous manga: after she joined Kodansha's Kiss, "My characters became much cuter," she observed. "Before working for Kiss, my drawing was, say, *Garo*-like," she said with a smile. By *Garo*, she referred to a legendary monthly manga magazine that featured alternative manga from the 1960s to 1990s. The artists whose works appeared in *Garo* had non-traditional drawing styles such as erotic, grotesque, avant-garde, and surrealistic types. Ito said, "Hers [her drawing style] is *maniakku* (maniac)," by which he implied a non-conforming, alternative style. One example of this style is the drawing of a character's bulging eyes, as shown with Bakka-chan in the left bottom panel of Fig. 5.3 (Vol. 2, p. 45). Okita added, "My [earlier] drawing style was *maniakku* and *sabukaru-teki* (subcultural). It was kind of a *kitanai* (messy; dirty) illustration."

I asked Okita to explain how Kiss made her drawing "cute." Looking at Ito, Okita answered, "My editor [Ito] told me to draw cute, to draw more reader-friendly." Because of that pressure, she said laughing, her characters ended up losing a nose. (See Bakka-chan in the bottom panel of Fig. 5.2 and the bottom right pane of Fig. 5.3.) A bit defensively, Ito added, "Well, what I asked for was something like 'How about drawing more neatly (*Mōchotto teinei ni kakimashō*)?'" Okita explained that Ito was not fond of the bulging eyes of her *Gaki-tame* characters. Usually accompanied with an *onyu* (sound effect) such as "*Gaan*" (Bang!), a bulging eye

Fig. 5.3 Image by Okita Bakka. (Courtesy of Kodansha)

is a gimmick commonly used in gag manga. Presumably, that sort of *deforume* (deformation) did not match the dominant *shōjo*-manga style of Kiss. Because I was not quite convinced, I asked them how cuteness relates to reader-friendliness. Ito replied that the cute drawing style "can cover a wider range of readership (*maguchi ga hirogaru*), after all."

Okita also stated that because she cannot draw elaborate backgrounds, she uses a lot of screentone patterns to make up for it. We all laughed. She then offered more insights into her art as follows:

> Screentone images are convenient, you know. There is one for thunder. It allows me to create the right atmosphere for *shuraba* (a moment of personal crisis). When I was drawing everything analog,[10] I used screentone sheets for background drawing all the time. I draw manga mostly with the basic *manga kigō* (signs used in manga semiotics), so basic that sometimes readers can't tell if the background is inside or outside or if the character is a child or adult. My drawing can be said to be a compilation of conventional manga signs (*kigō*).

Although she evaluates her artistry as "awfully messy drawing," the following examples show her creative ways in which she visually communicates the ins and outs of her neurological condition, making this manga a worthy piece for semiotic analysis.

Hypersensitivity

Since she was little, Okita has been inflicted with a heightened auditory perception, which is commonly called *chōkaku kabin* (auditory hypersensitivity) in the literature of *hattatsu shōgai*. When she was in elementary school, she could hear the lecture in the next room and could not concentrate on sounds in her immediate environment. Under such circumstances, her autistic brain can no longer absorb the lesson. I asked her how she attempted to visually represent her *hattatsu shōgai* symptoms in the medium of manga. Okita answered, "When writing *Gaki no tame iki*, I came to realize that my perception is drastically different from that of typically developing children, especially the way sounds are processed." As a child, her most noticeable symptom was *chōkaku kabin* (auditory hypersensitivity), which her teachers mistook for her overreacting or acting up.

To illustrate how auditory hypersensitivity affects one physically, Okita drew a scene in *Gaki-tame* in which Bakka-chan panics and vomits in reaction to the clanking sound of an aluminum lunchbox that causes a pain in her (from middle two panels to bottom panels in Fig. 5.4). Instead of consoling the crying Bakka-chan, her homeroom teacher scolds her for "overreacting" (see bottom left panel in Fig. 5.4). Another episode about her hypersensitivity depicts that when sitting near the window, she gets distracted by all the noises coming from the schoolyard where students are exercising in PE class (Vol. 1 pp. 58–59). Other episodes of *Gaki-tame* show her being tormented by the sound of chalk on the blackboard (Vol. 3 p. 59) or the clicking noise of another student's mechanical pen (Vol. 2 p. 32). "With these distractive noises, I couldn't concentrate on the lessons and appeared to be an absentminded child (*bōtoshita ko*). The teachers couldn't tell if I was paying attention to them at all," she said. She added that many children with *hattatsu shōgai* are hypersensitive to abrupt sounds such as the sounds of fireworks or the pistol at an athletic meet. She explained, "To certain sounds, I couldn't help responding with these troublesome behaviors. To describe in manga how my hypersensitivity resulted in a troublesome behavior, I drew all the despicable sounds (*iyana oto*) in a three-dimensional modality."

Here, Okita depicts how hypersensitive she was to sounds, drawing her mother's loud, angry words as they turn into letter blocks and hit her head (top two panels in Fig. 5.5). Pain was not the only reaction to certain sounds that her body experienced. On some occasions, she felt the sound's temperature: a "hot" or "cold" sound. On other occasions, she had a tickling sensation in her ears. Okita told me that when she shared this perception with a typically abled person, she was told, "You must be kidding. Sound is sound." That was when she realized that, to neurotypical individuals, a sound is something that only your ears absorb. Furthermore, Okita's brain perceives panic-triggering sounds as real, physical pain. To visualize that experience, she chose to draw these problematic noises as large objects that chase, hit, or tickle her. To show exactly how her body processed her mother's yelling voice, for example, she draws enlarged 3-D letters. "I feel I did a very good job on visually depicting this, although it was a bit experimental," she said. Okita believes that sort of strategy will help the parents of children with *hattatsu shōgai* better understand auditory hypersensitivity. That is her artistic way of informing them how sounds are perceived by some autistic brains and why certain sounds can scare these children. Okita said that by writing this manga, she gained a great deal of self-discovery. "It finally dawned on me that there are real reasons behind those behaviors. That realization came to me as I wrote this manga. To me, it was like 'A discovery, a discovery (*hakken da. hakken da*)!'"

59

Fig. 5.4 Image by Okita Bakka. (Courtesy of Kodansha)

Fig. 5.5 Image by Okita Bakka. (Courtesy of Kodansha)

To this day, Okita hates the sound of a diesel car. The sound of a motorbike used to freak her out, and she still dislikes the sound of a hairdryer. "I can't handle a sound that gets loud suddenly like 'Wooon!,'" she said. Only recently did she realize that she had avoided drying her hair for years because she hates the sound of the hairdryer. She shared her insight into a behavior that can be easily misunderstood as lazy or unkempt by saying, "Normally, if you go out with your hair wet, you give the impression that you are a very untidy woman (*sugoi darashinai onna*). Well, I wasn't aware of that impression either. Anyway, I'm actually not untidy but just unable to handle that sort of sound."

Hypersensitivity of one or more of the five senses (seeing, hearing, smelling, tasting, and touching) is very common among Aspies (e.g., Hirose 2015). Furthermore, children like Okita may also be hypersensitive to tastes, resulting in having strong likes and dislikes (e.g., those who drink only water and reject other beverages; those who eat noodles but not bread or rice). In one episode, Bakka-chan experiences semi-starvation because she cannot eat most of the food, particularly sweet beans, served at a temple camp (Volume 2 p. 71).

Poor Coordination

Children with *hattatsu shōgai* tend to have poor motor coordination as well. In one episode, Bakka-chan becomes the last child to learn to ride a bicycle in a school curriculum after her classmates completed the training, passed the final test riding, and received a certificate. She keeps flunking the test because of poor balancing skills and trouble following the instructor's directions such as "turn left at that corner" because she cannot immediately tell right from left (see Fig. 5.6: Volume 1 p. 97). In other episodes, she constantly bumps into objects as well as people surrounding her (e.g., Volume 2 p. 27). According to Hirose (2015), the difficulty of keeping good hand-eye coordination is one of characteristics shared by many Aspies, and in some cases, they even bump into a wall or a piece of furniture even in a spacious room, due to poor coordination (pp. 48–49).

"Writing about my school days allowed me to realize many things," Okita told me. For example, it helped her see why the teachers were upset at her at that time. "By revisiting those incidents [via writing *Gaki-tame*], things that I was clueless about back then are clear to me now as an adult. I tell myself, 'Oh, this is how they came to that conclusion and why they scolded me.'" The whole thing became transparent to her as a person

97

Fig. 5.6 Image by Okita Bakka. (Courtesy of Kodansha)

knowledgeable about this neurological condition. With *Gaki-tame*, Okita's objective was to document what sorts of behaviors got her in trouble with the teachers as well as her parents.

Selective Mutism and Suicide Attempt

As mentioned earlier, Okita developed *bamen kanmokushō* (selective mutism), or *kanmoku* for short, in the third grade. As this scene from Volume 3 (see Fig. 5.7) shows, that phenomenon occurs abruptly and then her speech inability dissipates as soon as an anxiety-triggering event is over. The moment in which Bakka-chan's mind sets that deadlock on is cleverly depicted with the *onyu* (sound effect) of "*bishi bishi*" (snapped) and screentone in the bottom panel of page 32. The character is covered entirely with the screentone filter when she is standing motionless with fear, but the moment everyone's attention leaves her, the screentone chips away in pieces as shown in the bottom left pane of page 33.

Although Okita completed high school and was successfully accepted for a vocational school, Okita's *hattatsu shōgai* symptoms worsened in

Fig. 5.7 Image by Okita Bakka. (Courtesy of Kodansha)

adulthood. In particular, she experienced communication difficulties at work and developed selective mutism (*bamen kanmoku*). "But back then, I couldn't explain why I fell silent. I knew I had to apologize at least, [for the trouble I had caused], but when you have *kanmoku*, you can't even say, 'I'm sorry,' you know. So I just clammed up," Okita described and paused briefly. *Gaki-tame* has an episode in which she is an adult and young nurse and attempts to take her life after her superior treated her with what appears to be power harassment (Vol. 1, p. 68). Okita explained how she felt as follows:

> Naturally, the others would get upset because a 20-something grown-up couldn't even say sorry. I was unable to understand what was going on myself, while a schism began to open between me and my coworkers. Even then, the idea that the cause was AS never occurred to me. I sensed that it was no longer a matter of personality; it was something more serious, but I couldn't make out what it was. Being unable to know how to solve it, I could do nothing but think 'I'm a useless human being' and ended up attempting suicide, which I survived.

She went on, "What was clear at that time was that I'm not fit to be a nurse." After she survived that suicide attempt, she figured that she would possibly last longer in a profession other than nursing. She searched for a less complicated job, which led her to the field of *fūzoku* (adult entertainment).

Before the interview, I had thought Okita's visual representations, especially with her choice of manga iconography, were hyperboles, typical of gag manga. However, her remark below convinced me that those depictions were not exaggerated for theatrical effects; they are, in fact, close renditions of her lived experience of *hattatsu shōgai* both as a little schoolgirl then and as an adult now. "I am in the subtype called 'ADHD *tadōsei yūseigata*,' so right now my head is busy processing a lot of noises (*taihen atama no naka ga urusai*). If I tell people about it, they say, 'Our heads aren't like that.' That makes me wonder, Okay, 'normal' people don't hear music in their head when they are just walking, do they? I usually have the music of Super Mario playing in my head. But typically abled people don't hear such a thing. That's just one example that my brain processes things differently than the brains of others. So, I was determined to depict those things in *Gaki no tameiki*. Sure, manga as a medium is prone to exaggeration. But more or less, my drawing is true of my experience."

Ninchi no Zure *(Cognitive Gap)*

Okita confided, "I cannot get emotionally connected to others (*kyōkan dekinai*). This lack of ability to empathize with others—this is a bit punishing (*shindoi*)." Attributing that inability to her *hattatsu shōgai*, she said, "I don't always feel sad when others are mourning or funny when others are cracking up." Because she could not demonstrate the same emotional reactions such as laughing with others, she practiced how to laugh "normally" in elementary school by watching a TV program run by *Dorifutāzu* (The Drifters). Known for their slapstick comedy, this popular music and comedy band had several TV variety shows including *Hachiji dayo! Zeiin shūgō,* or "It's 8 O'clock! Everybody Gather 'Round!" in the 1980s and 1990s. That practice served as self-coaching for her to learn when and how people laugh. "As a consequence, I now burst into a horselaugh, and my neighbors complain that is too loud," she said jokingly.

People with *hattatsu shōgai* are said to have *ninchi no zure*, or cognitive gap, which is a tendency to interpret and act in situations differently from neurotypical individuals due to their unique brain functions (e.g., Hirose 2015). For example, Aspies tend to take a literal understanding of what others say, missing connotations and implications involved in the statement. They may also have difficulty inferring others' emotions from their facial expressions and other body language cues and empathizing with them. Although Okita felt there was "something wrong with me," she did not understand why stepping out of group mentality (*shūdan kōdō kara hazureru*) was such a big deal to others, especially to teachers. Sometimes, teachers made her stand in class for punishment; other times, she was put in a kangaroo court (*tsurushiage raretari*) in a homeroom meeting. The types of punishment she experienced were corporal or psychological. In reference to her LD, she said, "So I was deemed slow (*atamaga sugoku warukute*). Teachers were upset with me because, in spite of being a slow student [who should be compliant and seek their help], I didn't obey them. They probably thought, 'Who the hell is this child?'" After learning that Bakka-chan had a disability, her mother encouraged her to enroll in a rehabilitation center for autistic children and join a support group for children with LD (see Fig. 5.8). That is one of only a few heartwarming scenes of her memoir, however.

On reflection, the grown-up Okita understands now that those teachers must have been equally confused and distressed over the situation because they had no knowledge of *hattatsu shōgai.* Indeed, when she was in

156

Fig. 5.8 Image by Okita Bakka. (Courtesy of Kodansha)

elementary school some thirty years ago, the public had no awareness of *hattatsu shōgai* in Japan. "My third-grade homeroom teacher, who was a woman, had a really hard time dealing with me. That teacher could not communicate well with me even though she attempted different ways to connect with me." Being unable to focus on teaching and class management, the homeroom teacher encountered *gakkyū hōkai* (classroom chaos). That was when she literally "hit the wall (*kokoro ga oreru*)" and resigned from the school. Okita pointed out, "Of course, I wasn't the only reason for the teacher's resignation." However, this episode, which appears in *Gaki-tame*, demonstrates how a single case of *hattatsu shōgai* can go awry and affect everybody else if no support is available.

Reception and Generalizability

Unlike her *Tomei na yurikago* series, *Gaki-tame* has not accrued any award formally. However, the publication of Okita's *tōjisha* manga, including *Gaki-tame*, helped increase her visibility in the media. For example, NHK, featured Okita as a "manga artist with *hattatsu shōgai*" (https://www.nhk.or.jp/heart-net/program/heart-net/1186/) in 2019. I told Okita that, if more people read her *tōjisha* manga about childhood and adulthood *hattatsu shōgai*, society would become more aware of this condition, and possibly, the entire nation would make more effort to mitigate the *ikizurasa* (difficulties of living) of child and adult Aspies. In response to my speculation, she said, "I heard that schools have my manga available for students' reading." According to Okita, schools of nursing (*kango gakkō*) and universities of medicine, in particular, keep her manga as "*shitei tosho*" (recommended readings). "In that regard, I can see my manga has made some contribution to education," she said. As mentioned in chapter "Tōjisha Narratives", such readings are called "*narrative kyōzai*" (disability memoirs used as educational material).

Okita has given a lecture at a university titled "*narrative kyōzai*". She explained, "Because those are writings by the *tōjisha*, many readers find them very intriguing. Even college professors and specialists don't know much about this. These writings offer real-life data and help them learn more about *hattatsu shōgai*. I was told that my manga has been very helpful in that regard." Recently, she has also been engaging in *taidan* (face-to-face interviews) with various people about this condition. From those *taidan* experiences, she came to realize that "doing something in a large number of same-age peers to achieve the same level of success—that type of act itself is already out of fashion." Children who cannot perform accordingly will end up being *futōkō*, or truant, a consequence that still carries a negative image in society. Okita asserts, "But it is the school culture that is failing them, labeling them as losers."

She emphasizes that her tōjisha narrative does not necessarily account for everybody's situation, saying "What I write about is strictly my own story derived from who I am as an individual (*akumade watashi kojin no seikaku kara kuru tteiuyōna*)." Although she understands the diverse backgrounds of *hattatsu shōgai* tōjisha, she still makes use of her unique experiences to advocate for Aspies. She told me that when writing the third volume of *Yarakashi*, she came to realize that many adults are out there struggling with that condition and that she has to be a voice for them. One memorable interview that she conducted as an advocate is one she had with a guest speaker nicknamed "Ogi Mama." According to Okita, one of her manga publishers "hooked me up with him." For a special broadcast on *hattatsu shōgai*, Okita was asked to interview him. Ogi Mama is a nickname of Ogi Naoki, a professor emeritus of Hosei University and an education commentator (*kyōiku hyōronka*,) who frequently appears on TV. Having published more than 200 publications since the 1970s, Ogi is known for his opinions about *ijime* countermeasures. With a unique speech mannerism and outfit, he has reached the status of a celebrity in the media, and young audiences fill his talk events. Interestingly, Ogi and Okita came to the same conclusion that we must eradicate the problem of school *ijime* (bullying), targeting children who are perceived as "the other." As she told him her story, it occurred to her that the school's health room (*hokenshitu*) might be an ideal place for one-on-one counseling for children with *hattatsu shōgai*. They both believe that Japan "needs to adopt a Western style" because their educational system has operated under an insular ideology in which individualism crashes with collectivism. "We agreed that we should look beyond our country [for solutions]," she said. She emphasized the need for diversity in education where "If you can't read, it's accepted as a trait of *hattatsu shōgai*. Nothing more, nothing less." She described Japan as a country where a disability is treated rigidly.

What has concerned her most recently is that although more people know about *hattatsu shōgai* nowadays, wrong information about the condition has also been disseminated. For example, some people infer that having *hattatsu shōgai* equals being a genius. Relating this phenomenon from her own childhood, she said, "My mother was like that, believing this child has something phenomenal in her." Because of Okita's hypersensitivity to sounds, she incorrectly thought it was a sign that the little Okita was destined to become a pianist and sent her daughter to piano lessons. But Okita could not read musical notes and simply imitated other children. She said, "But we must know that not all of us can become Steve Jobs. [Laugh]. Only a fraction of us can be." Therefore, she cautions against stereotyping the condition as a "genius" badge, saying, "Because of that misconception, some mothers get anxious about finding a 'special talent' their otherwise underperforming children are supposed to have."

Toward the end of the interview, I asked Okita if she considers manga better suited to illustrate her neurological disability than film or novels and what this medium allows her to do that the other media do not. Hoping to hear Okita's opinions about manga's strengths and weaknesses, I was a bit baffled when she replied rhetorically, "Is there such a thing?" However, soon she added, "I use my editor's opinions for reference quite often. Basically, I draw without thinking consciously." Okita seems to value the process of her editor looking at her penciled storyboard (*nēmu*) and providing her with feedback. She explained to me that particular process as follows:

> I bring a piece I drafted with much enthusiasm, but it can be rejected right away. [Laugh] At the beginning [of my career], I used to be really upset when I was told, "This is not right." I protested [to the editor], "Why do you reject the storyboard I worked so hard on for days?" But my husband said, "That's because it wasn't interesting." Then, I was like, "Oh, I get it." [Laugh] My husband then said, "It'd be a waste of your time to get upset about a draft that is uninteresting anyway. Use that time to think of something new instead." He is older. Even though his storyboard is criticized severely and he does get upset, he accepts it [the criticism] the next day, saying, "It was my fault." His thinking is that the editor's job is to terminate/reject (*kiru*) the storyboard that is not appealing to readers, and that's my fault, not the editor's. So he advised me to do the same, saying, "It's useless to talk back like that." I said, "Yes, sir!"

Okita told me that she does not trust her original ideas and she does not rely on just herself to judge "what is good." Laughing with Ito, Okita reasoned, "I actually draft my drawing on the assumption that I will get scolded. My mind is conditioned to think my drawing will get critiqued by my editor or someone else [in the publisher's office]." She has learned that critiquing is part of the process and that the stuff they approve is probably what readers will enjoy reading. During the interview, I noticed Okita had formed a symbiotic relationship with Ito, who has worked with Okita throughout the *Gaki-tame* series and has become a good supporter of this manga artist. Ito also appears in Okita's manga (e.g., *Gaki-tame*, *Yarakashi*). Their amicable work relationship is the antithesis of the tragic case of Yamada Hanako and her *tantō henshūsha* (Ishikawa 2001). Even though she does not articulate the strengths of her artistic styles, it is undeniable that with her supportive editor, she has produced autobiographical comics that help readers learn about *hattatsu shōgai* by making the invisible brain functions visible through the medium of manga.

Notes

1. In Japan, mental disorders are typically divided into two large clusters: *hattatsu shōgai* (developmental disorders), a category that includes autism spectrum disorder, ADHD, and Learning Disorder (LD), and mental disabilities, a group that includes depression, schizophrenia, mood disorders, and substance-related/addictive disorders. While DSM-5 classifies autism spectrum disorder in the class of neurodevelopmental disorders, ICD-11 places it in the section of mental, behavioral, and neurodevelopmental conditions along with mental health conditions with psychosis.
2. Besides these memoirs about her *hattatsu shōgai*, she has published an autobiographical manga, *Shinkirō Kazoku* (*The Mirage Family*), featuring Okita's family, which she said she had written "without thinking carefully" and was almost sued by one of the families, especially because she used the people's real names in the story. (Both Okita and Ito laughed, however, indicating that the situation was not so serious.) With its story, *Shinkirō Kazoku* is considered a gag manga, and Volumes 1–5 were published in 2015–2016. Okita said, jokingly, "Because of that, I'm a *hotte okuto chotto nanisuruka wakannai mitaina* (unreliable person who needs some supervision). I write whatever comes to my mind. So I need someone to stop that."
3. In Japanese, "breathing" is *iki wo suru*, which shares the same word *iki* (breath) with *tameiki*.
4. The word *ryōiku* refers to a type of special training for children with *hattatsu shōgai*; its process involves both the parent and autistic child (Kondo 2018). In Japan, one- to two-year-old toddlers can be tested for Asperger's syndrome. An autistic baby typically has trouble falling asleep and tends to wake up frequently. Thus, it is recommended that such a baby be examined right away. By the age of two or three, ASD can be confirmed by means of the child's biorhythms. After that early diagnosis, the next step is providing

the child with *ryōiku*, or rehabilitation education for children with *hattatsu shōgai*, in facilities similar to Asperger's treatment centers in the United States that provide educational therapy for autistic children. If *ryōiku* is provided before entrance to elementary school, the child is likely to have learned to deal with various social situations, even somewhat difficult ones. Then, by the time of adulthood, the person's *ikizurasa* (difficulties of living) will be greatly reduced. The official enforcement of *ryōiku* varies from prefecture to prefecture, however. Some prefectures run a campaign such as "Let's protect children with *hattatsu shōgai*!" while others are totally opposite.

5. A genetic aspect of *hattatsu shōgai* (especially ASD) has been identified (e.g., Hirose 2015; Iwanami 2017). For English language-based research on this topic, see Huguet et al.'s (2016) "The Genetics of Autism Spectrum Disorders," for instance.
6. *Hikikomori* can be total or partial isolation at home and withdrawal from school and society.
7. Japan issues a *shōgai techō* (disability passport) to those who have disabilities severe enough to prevent them from working. The Municipal Office of Tokyo issues a *shōgai techō* specifically for children with *hattatsu shōgai*, and the unique name of the passport is "*Ai no techō*" (literally, "Passport of Love"), while the standard name of the passport is the *ryōiku techō* ("rehabilitation passport"). Regardless of the name, the government's "love" seems rather conditional, as only children with *hattatsu shōgai* whose IQ is below 70 are officially qualified to have that passport. Aspies whose IQ is higher than 70, so-called Borders, are not qualified for that passport. If the person has no intellectual disability, he/she can only apply for the other type of disability passport for mental disabilities: *seishin shōgaisha hoken fukushi techō*. However, this passport is designed for adults diagnosed with *hattatsu shōgai* later in life; they can apply for this passport when trying to find a job at a company that has a number of positions set aside for disabled individuals (*shōgai waku*). Thus, this passport does not enable school-age children to receive any benefits and support.
8. Mixi was an old social media site in Japan, a sort of a precursor to Twitter or Facebook, and was active particularly in the 2010s. It allowed users to create their homepages easily.
9. The term "*ofu-kai*" literally means "off-meeting," used in reference to face-to-face meetings with members of an online community in Japan.
10. In the manga industry, "drawing analog" means drawing manually with pen and paper only—the traditional method of using a *tsuke* pen (a pen and ink bottle) and *manga-yō genkōyōshi* (manuscript paper designed for manga), while "digital" means drawing with computer software, the most common approach to manga drawing nowadays.

References

DSM-5 (*The Diagnostic and Statistical Manual of Mental Disorders*). 2013. Washington, DC: The American Psychology Association.

Hirose, Hiroyuki. 2015. "*Moshikashite Asuperugā?*" *to omottara yomu hon* (*A Book to Read When You Suspect Asperger's*). Tokyo: Nagaoka Shoten.

Huguet, Guillaume, Marion Benabou, and Thomas Bourgeron. 2016. "The Genetics of Autism Spectrum Disorders." In *A Time for Metabolism and Hormones*, edited by Paolo Sassone-Corsi and Yves Christen, pp. 101–129. New York: Springer.

Ishikawa, Gen. 2001. *Kakusareta shōgai: mangaka Yamada Hanako to higengosei LD* (*Concealed disability: Manga Artist Yamada Hanako and Her Nonlinguistic LD*). Tokyo: Iwanami shoten.

Iwanami, Akira. 2017. *Hattatsu shōgai*. Tokyo: Bungei Shinju.

Kayama, Misa, and Wendy Haight. 2014. *Disability, Culture, and Development: A Case Study of Japanese Children at School*. New York: Oxford University.

Kondo, Naoko. 2018. *Ryōiku tte nani?* (*What is ryōiku?*). Tokyo: Kurieitsu Kamogawa.

Okita, Bakka. 2013a. *Nitoro-chan* (*Little Nitoro*). Tokyo: Kobunsha.

———. 2013b. *Mainichi yarakashitemasu* (*Everyday life of an adult Aspie*). *Volume 1*. Tokyo: Bunkasha.

———. 2015. *Tomei na yurikago: Sanfujinka-iin kangoshi minarai nikki* (*An Invisible Cradle: A diary of a nurse-trainee at an OB-GYN clinic*). *Volume 1*. Tokyo: Kodansha.

Tateiwa, Shinya. 2014. *Jisheishō renzokutai no jidai* (*The Era of Autism Spectrum Disorder*). Tokyo: Misuzushobo.

CHAPTER 6

Nonami Tsuna's *Akira-san* Series (2011–2017): A Memoir of a "Cassandra" Wife

Introduction

Nonami Tsuna's *Husband, Akira, Has Asperger's Syndrome* is an eight-volume series published by Kosumikku Shuppan (Cosmic Publishing) (Fig. 6.1). These volumes were released over a span of seven years.[1] After the first volume, the title of the rest begins with "*Akira-san wa asuperugā*" (2011) followed by a subtitle. For instance, Volume 5 is titled "*Husband, Akira, Has Asperger's Syndrome, Wife Has Cassandra*" (*Akirasan wa asuperugā, okusan wa kasandora* 2014), Volume 6, "*Husband, Akira, Has Asperger's Syndrome, Asperger's and Cassandra*" (*Akirasan wa asuperugā, Asuperugā to kasandora* 2015), and Volume 7, "*Husband, Akira, Has Asperger's Syndrome, Let's Find Hints of Happiness*" (*Akirasan wa asuperugā, Mitsukeyō egao no hinto* 2016a) in 2016. Unlike the other volumes, the final Volume 8 is a summary narrative in which Nonami reflects upon her courtship with Akira first and then their marriage, parenthood, and eventual separation. Hereafter, the abbreviated title, "*Akira-san* series," will be used to refer to the entire series or any part of it.

Using this autobiographical manga, this chapter briefly introduces readers to husband Akira's "adult Asperger's syndrome" (*otona no asuperugā*), a popular phrase in Japan nowadays, and then focuses primarily on wife Tsuna's Cassandra syndrome (*Kasandora shōkōgun,* also known as *kasandora jōdō hakudatsu shōgai*). Cassandra syndrome, or Cassandra, for short, is a mental health phenomenon experienced primarily by women

Y. Okuyama, *Tōjisha Manga*,
https://doi.org/10.1007/978-3-031-00840-5_6

Fig. 6.1 Image by Nonami Tsuna. (Courtesy of Kosumikku Shuppan)

who are trapped in a close relationship with a sufferer of Asperger's syndrome.[2] In the United States, Cassandra syndrome is commonly referred to as the "Cassandra Phenomenon" or "Affective Deprivation Disorder." In my interview with her, Nonami described her painful experience of this condition, and her *tantō henshūsha* (manga editor), Amano Kyoko, who accompanied her, occasionally offered her insightful comments as someone who works in Japan's comic industry. Nonami told me about how her mental health was profoundly affected by her husband's neurological condition, leading to her development of this debilitating health issue. Throughout the interview, she emphasized how publicizing this mental health fallout is relevant to and equally important for public understanding of Asperger's syndrome.

It is likely that the term *Cassandra Syndrome* began to be mentioned in Japanese mass media around 2014. For example, the term was used in Shūkan Asahi's 2014 article, "Wives Who Live with AS Husbands: Do You Know 'Cassandra Syndrome'?" and in *Yomiuri Newspaper*'s blog for women, *Hatsugen Komachi* (https://komachi.yomiuri.co.jp/t/2014/0507/656971.htm) in 2014. In the same year, the first volume of the *Akira-san* series was published as a *tankōbon* (collected edition of manga). Nonami uses the word "Cassandra" (*kasandora*), not "Cassandra syndrome," in the *Akira-san* series partly because Cassandra syndrome is not recognized as a disorder in either ICD-11 or *DSM-5*, although the term is often used in counseling to refer to those mental health conditions commonly suffered by the neurotypical spouses. By sticking to the use of "Cassandra," Nonami intends to avoid creating an impression that it is the established name of a disease. Volume 6, in particular, illustrates how wives of Aspies, commonly known as "spouses of Aspies" (*asuperugā haigūsha*), are distressed by communication difficulties and relationship troubles, feel isolated from a lack of connection with their partners and social support from those close to them, and eventually develop their own psychological disorders such as depression and anxiety disorders. The volume depicts how the character Tsuna's emotional appeal for help was dismissed as a minor complaint by others, including the people close to her.

Akira's Asperger's syndrome is the main focus of Volumes 1 through 4 that depict hilarious episodes about Akira's condition. To be clear, Asperger's syndrome (*Asuperugā* or *Asuperugā shōkōgun* in Japanese) is a neurological condition, not typically considered a mental illness. Asperger's syndrome is not associated with intellectual disabilities, nor does it entail delay in language learning or cognitive development (Kayama and Haight 2014). People with this neurologically-based disorder have difficulty with verbal and nonverbal communication such as maintaining eye contact and understanding abstract speech such as metaphors, euphemism, and sarcasm; they also tend to demonstrate an intense interest in certain topics and objects and have strict adherence to routines, preferring repetitive movements; they may experience difficulty maintaining occupational or other social relationships with peers; some of them lack interest in participating in social activities or showing empathy to others. Because there is no delay in speech development and other symptoms may be relatively mild, it is quite common that the condition is not diagnosed until adulthood, as in Akira's case. In addition to these communication and relationship difficulties, a "unique" sense of budgeting is possessed by some Aspies (Hirose 2015), as in the case of Akira.

In Volumes 5 through Seven, Nonami imparts her own suffering of Cassandra syndrome as well as those of other women married to men with AS. She told me that for the first four volumes, she had prioritized her need to explain what behaviors characterize Asperger's syndrome in the first place while suppressing a detailed description of her own feelings because, back then, the public was generally unfamiliar with the term *asuperugā* (Asperger's). For this chapter, I primarily analyze Volume 1, which introduces the main characters and basic storyline, and Volumes 5 through 7, which describe Cassandra syndrome in detail, to contextualize Nonami's mental health experiences. The reason for weighing more on the latter volumes is that she shared her perspective as a Cassandra *tōjisha* in the interview, and her lived experience of Cassandra syndrome is highlighted in those volumes.

In this chapter, the term "Cassandra syndrome" (and sometimes its shorter version, "Cassandra") is used consistently, unless the exact verbatim is drawn from the interviewee's Japanese speech. Furthermore, the manga artist is referred to as Nonami, and her character, as Tsuna.

The Title and Target Readership

The original Japanese title of *Akira-san wa asuperugā* is 旦那さんはアスペルガー (*Husband, Akira, Has Asperger's Syndrome*). The titular character, Akira-san, is Nonami's real-life husband. It is important to point out that although the title uses the *kanji* characters 旦那 (husband), the *furigana* (pronunciation guide) written above the *kanji* reads アキラ (Akira), implying that the story centers around the role of Akira as a husband and illustrates how his condition impacted their marriage and family life. Nonami also draws herself as the wife Tsuna who serves as the narrator of the story, reporting on their lives from Nonami's point of view. In the manga, she uses the characters' first names only. Nonami had a daughter and a son with Akira, and they had lived together until Akira's secret debt grew immense, compelling them to sell their house, both in the series and real life.[3]

Nonami told me that a psychiatrist, Dr. Miyao Masutomo, diagnosed Akira with Asperger's syndrome in 2010. Dr. Miyao, who oversaw the entire *Akira-san* series, appeared and offered insights as a medical consultant in the first five volumes and in Volume 8. Dr. Miyao runs a clinic to treat young patients with autism spectrum syndrome including Asperger's and has published several books including *A Book to Solve Cassandra Mothers' Distress* (2018). Takiguchi Nozomi, a counselor who specializes in couples therapy for Aspies[4] and their spouses, also commented in Volumes 5 through 7, since those volumes highlight Nonami's symptoms from Cassandra syndrome in particular. Takiguchi is a licensed counselor who works at Dr. Miyao's clinic. Although Miyao or Takiguchi's name might be inadvertently shown as co-author in some search engines (e.g., WorldCat.org), Nonami is the sole author of the series, creating both text and graphics by herself.

When Nonami began writing essay manga, the genre mostly covered topics in the course of women's lives, like marriage and motherhood. Back then, she was writing about cats and other slice-of-life stories and would never have imagined writing a story about such a serious topic as mental health. However, while working productively in this genre, she began to realize that something was wrong with her marriage. At that time, she had no idea that Akira had Asperger's syndrome (AS). His AS-related behavior began to negatively affect her health, so much so that it also interfered with her work. Even though she tried hard to write funny, lighthearted stories, she was unable to produce even a single work while preoccupied with the problems at home. Finally, realizing that she had reached an impasse, she took a hard look at who her husband was and why he behaved the way he did.

Decades before Akira's diagnosis in 2010, people had only heard of *jiheishō* (autism) in Japan, a term associated with the image of an "autistic savant" with intellectual disabilities such as the protagonist of the 1988 movie, *Rain Man*, who had severe impairments in speech and self-care. However, after the 1981 publication of Lorna Gladys Wing's seminal paper about Hans Asperger's research, the understanding of Asperger's syndrome began to spread worldwide. (For more information on autism, Wing, and Hans Asperger, see Okuyama 2020). Although books initially focused on childhood autism in the 1970s and 1980s, the terms *Asperger's* and *autism spectrum disorder* began to appear in the titles of publications by mental health specialists in the 1990s and 2000s in Japan (e.g., *Anataga anata de arutameni*, "To Remain Who You Are," by Yoshida Tomoko, 2005). Therefore, when Nonami sought information about Akira's condition to understand their family dynamics, these publications did not help, as they were mainly written for educators and medical professionals who treated the tōjisha directly, not their families and relatives.

Nonami told me that before her manga was published, most books written about Asperger's syndrome in Japan informed readers mainly about how to support the *tōjisha*. She said, "The books only tell you this is how you can get behind Aspies." She felt that some wives may be telling themselves, "I must wait on my husband hand and foot (*sasae nakya*)." "I was like that too," she said. While acknowledging the importance of family support, Nonami argues that these books understate the difficulty of supporting the *tōjisha* singlehandedly. She felt that neurotypical spouses are left to struggle alone and wanted to tell others how mentally and physically demanding it can be to live with Aspies. Wanting to share her feelings of frustration, anger, loneliness, and guilt with other struggling women, she decided to write about her experiences so that she could make her readers more aware of how difficult it is to live with an Aspie because, in her view, "a wife is responsible for doing more than simply providing her husband with emotional support." She emphasized, "A professional caretaker can leave after her work is done. But a spouse is stuck there."

In the summer of 2010, Nonami came up with the idea of making manga out of her experience as the wife of an Aspie. That year, she proposed the project to Cosmic Publishing.[5] The first volume of the *Akira-san* series was published in 2011. Although the term Cassandra syndrome was introduced in Volume 4, it was not until Volumes 5 (2014) and 6 (2015) that the series' central theme turns to this psychological condition that is likely to be developed by neurotypical spouses of Aspies. These volumes highlight her niche contribution not only to the genre of essay manga but also to the field of publications that promotes mental health awareness.

What motivated Nonami to switch her attention from Asperger's to Cassandra syndrome? As the *Akira-san* series continued to receive a flood of audience responses from fans, mostly female readers, Nonami felt it necessary to focus on Tsuna's feelings and thoughts regarding her Cassandra syndrome in Volumes 5, 6, and 7. In those volumes, in particular, Tsuna shows the hardship of cohabitating with someone with Asperger's syndrome when she has no knowledge of the condition herself. In other words, unlike the earlier volumes which focused on Akira's symptoms, the later volumes zero in on the psychological and physiological impact of his mental disorder on Tsuna. Nonami eventually lives separately from her husband and her symptoms gradually disappeared. In Volume 3, Akira is shown to begin living independently for the first time in his life and support himself by working as a security guard, after the family is forced to sell their home because of Akira's debts. In Volume 5, Tsuna is depicted to feel as though she abandoned her disabled husband. In our interview, Nonami also confided in me that she still could not shake the sense of guilt totally.

Volume 8 is titled "*Husband, Akira, Has Asperger's Syndrome, A Marriage without the Knowledge of Asperger's*" (*Akirasan wa asuperugā, Asuperugā to shiranaide kekkonshitara tondemonaikoto ni narimashita*) and was published in 2017. It is the final volume of this series born from Nonami's reaction to the questions she most often received from both her readers and her friends. For example, she was asked, "Didn't you know that Akira had Asperger's syndrome before marrying him?" Some people who had never heard of Asperger's syndrome also asked, "Didn't you sense that there was something wrong with him?" This was often followed by a comment such as "It's your fault that you married such a person." The most frequently asked question was "Why didn't you notice anything wrong with him before marrying him?" As a response to these questions, Nonami decided to create Volume 8 to depict the period of courtship before she and Akira married. The volume first shows that Akira appeared to be an ideal suitor before marriage, explores the reason that Tsuna fell in love with him, and then describes how the change in roles from partners to parents affected the way he behaved. Volume 8 also illustrates how Akira began to reveal his true colors, which either he never did earlier or she never noticed when they were lovers, as well as how Tsuna learned that Akira expected "his" home to be where he could do whatever he wanted, an idea that sometimes conflicted with his responsibilities as a father. Believing that manga is the best way to respond to these questions,

Nonami hopes that Volume 8 helps readers to understand what it is like to be married to someone with Asperger's syndrome.

Nonami's target readers (in her words, "*ichiban yonde hoshikatta dokushasō*") are not Aspies but instead their spouses and other intimate partners.[6] She had this readership in mind from the beginning because, with the *Akira-san* series, she aimed at reaching out to neurotypical partners, mainly wives, of men whose Asperger's syndrome is taking a toll on their relationship. In our interview, she emphasized that ever since the first volume, her intention had been to share her experiences publicly for the sake of this particular group of wives. Nonami chose this target readership because she sees the issue of living with Asperger's syndrome as distinctively "a gender issue." At first, I associated her gender reference with what Goffman (1963) calls "courtesy stigma": prejudice and discrimination experienced by intimates of a person whose social value is perceived negatively, due to his or her "undesired" differences. As Jegatheesan (2009) states, courtesy stigma is experienced especially by mothers in Asian countries, because they are more likely to be blamed for bringing to life and raising "less-than-perfect" members of the society. It is a common phenomenon in many Asian cultures where the community highly values children's intelligence and giftedness. However, Nonami is Akira's wife, not his mother.

It turned out that Nonami was not referring to the social stigma imposed on her. Instead, she told me that she had learned from other wives of Aspies that many women feel pressured to be good wives by altruistically supporting their husbands and shoulder more of the household responsibilities than their husbands, along with the stressful demands of dealing with their husbands' Asperger's symptoms. This "gender issue" reminded me of a Japanese saying, *naijo no kō* (the wife's assistance), an idea that behind every successful man, there is his wife, who takes care of the home. During our interview, Nonami remarked, "It's been my hope that the series is read by the wives who are in the same boat as I was. By reading this manga, they may get it and say, 'Oh, this is my husband!' I wanted the manga to help those women become aware of that."

The crux of the matter is that in many instances, a husband with Asperger's syndrome is not cognizant of his own mental condition, nor does his neurotypical wife know much about the disorder. Under such circumstances, the wife receives little or no sympathy or thanks for her efforts and patience. In the interview, Nonami spoke out saying, "It's quite a lonely existence." She believes that if the spouse's brain problem were a clearly visible condition like dementia, people would be far more sympathetic to wives who live with their mentally declining husbands; the wives would be showered with praise and sympathy, as well as comments like, "You are doing a great job!" and "It must be tough. Thank you for your hard work." Instead, she went on, "Wives of Aspies are more likely to be perceived as selfish complainers if they speak of their problems." Nonami was often told, "You should be thankful that your husband is providing for your family." She then thought to herself, "Is that all I get for the hard work I do?" No one takes neurotypical spouses seriously, and no one acknowledges their efforts. Such an acute sense of isolation largely contributes to the development of Cassandra syndrome.

The term *Cassandra syndrome* began to attract public attention in the 2010s, although this mental health condition is not included in the *DSM-5*. Cassandra syndrome describes the neurotypical spouse's isolation and struggle for validation and support from extended family and friends. Cassandra syndrome is believed to be strongly associated with the stressful living conditions derived from being in a close relationship with an Aspie. In the *Akira-san* series, Nonami defines Cassandra syndrome (*Kasandora shōkōgun*) as "the condition of having psychological and physiological disturbances due to their partners' communication difficulties, one of the characteristics of Asperger's syndrome, which was identified by Aston, a British marriage counselor" (Vol. 7, p. 12). Maxine Aston is a British counselor who specializes in therapy for couples and provides workshops on relationships involving people with Asperger's syndrome. She has published books on the topic (2001, 2003, 2009/2020).[7]

The name "Cassandra syndrome" is derived from Greek mythology. Cassandra, a princess of Troy, was given the ability to foresee the future by the Olympian deity Apollo. Because Cassandra foresaw that Apollo would eventually abandon her, she refused his romantic advances. The angered god put a curse on her so that nobody would take her prophecies seriously. Thus, to her dismay, she was unable to convince others of the danger and alter the course of Troy's tragic fate. Similar to the circumstances of Cassandra, neurotypical spouses are rarely taken seriously by others when they appeal to their difficulty in living with their AS partners. Cassandra syndrome does not exist by itself; the condition develops only in reaction to Asperger's syndrome. That is why there is no therapy to treat Cassandra syndrome alone; the counselor needs to treat the relationship between the patient and their Aspie partner. The therapy aims at helping their relationship work. Both Aston (2009/2020) and Okada (2018) write that the couple must work together to understand their different communication needs and styles and to develop better ways of connecting with each other. Only then can couples counseling be effective. In other words, unless both partners are willing to work, therapy alone cannot heal Cassandra syndrome.

For a therapeutic intervention to be effective, however, both the Aspie and his neurotypical partner must be involved in couples counseling. Nonami told me that Akira had no motivation for counseling, and when he was diagnosed with Asperger's syndrome, he simply reacted by saying, "Oh, is that so?" Looking a little frustrated, she said, "That was the end of it." There is a type of training such as how to communicate and so on designed for adults with Asperger's syndrome. However, Akira said to her, "I'm not having any trouble with it, so I'm not interested in the training [designed for adults with Asperger's syndrome]." From Nonami's point of view, Aspies without awareness of their problems (*jikaku ga nai tōjisha*) are the most troublesome. Because they do not realize that their condition is troubling others, especially their family members, they do not acknowledge that their spouses are trying so hard to accommodate them. She said, "Without that acknowledgement, we'd wonder what are we doing all this for? It just feels like our efforts are fruitless." It is not uncommon that the untreated Aspie and his neurotypical wife decide to go separate ways eventually, as in Nonami's case.

As Aston (2009/2020) explains in her revised edition of *The Asperger Couple's Workbook*, living with an Aspie causes his neurotypical partner poor self-esteem, frustration, rage, and other emotional distress, and in some cases, depression, especially if an external support system is absent. When those who are closest are not empathetic enough, Aston argues, the neurotypical partner experiences emotional deprivation and acute loneliness. It is, therefore, precisely to those wives who are going through "emotional deprivation and loneliness" that the *Akira-san* series is dedicated.

The Author's Background

Nonami Tsuna is her penname,[8] and her *mangaka* (manga artist) career took off in 1989. Thus, by the time I interviewed her in 2019, she had been a *mangaka* for more than thirty-one years. After dropping out of Tama Art University (*Tama Bijutsu Daigaku*), she first acquired manga drawing skills by working as an assistant for an artist of a *shōjo* manga (comic aimed at young female readers). Soon, Nonami chose to draw *seinen* manga[9] and submitted her debut work to a *seinen* comics magazine because that genre seemed in more demand at that time. However, as a young twenty-something artist, Nonami began to see a schism between what the target male audience of *seinen* comics wanted and her own drawing style. (Back then, a popular *seinen* manga series was Masamune Shirow's *Ghost in the Shell* (1991), serialized in Kodansha's *Weekly Young Magazine* from 1989 to1990, which featured voluptuous female characters as ideal heroines.) The idolized, sensual, female protagonist typically drawn by these male *seinen* manga artists was not what Nonami wanted to draw, and the mismatch resulted in the eventual end of her series. While she was wondering what to do next, one of her friends suggested, "Why don't you write comics for a ladies comics (a Japanese term for "adult women's comics") magazine? With the growing number of ladies comics magazines, there is a labor shortage of manga artists in that genre." This suggestion motivated her to switch genres again, this time, from *seinen* to ladies comics.

Later, Nonami married Akira, an editor at a corporation that managed the editorial work of various publishers, and subsequently became pregnant. Before her pregnancy, she wrote ladies comics that were considered risqué, such as a story in which a woman under extreme stress commits a crime. But after her maternity leave, she felt that she could not relate to that sort of story anymore. The experience of giving birth made her unable to stomach malicious stories; instead, she desired peace. Realizing that ladies comics was no longer her preferred genre, she took a leave of absence. Through Akira's introduction, she found a new career path in writing "slice-of-life" (*nichijō*) manga, a genre now commonly known as "essay manga" (*essei* manga).[10]

It just so happened that her first offer to write essay manga came through that connection. She also chose the genre of essay manga because it appeared to be in greater demand by readers, and writing about how goofy and comical she was as the protagonist also fit her storytelling style. Her manga stories were well received. In addition to the *Akira-san* series, Nonami has published two other major works, an essay co-authored with Yagishita Noriko (2013) and *Our Mischief Maker* ("*Uchi no komatta-san*" published by Riido, 2016b). In our interview, Nonami (2019) pinpointed three representative characteristics of the essay manga genre: its predominantly female readership; coverage of topics that are not serious social issues but instead are common experiences, especially among women (from shopping and traveling to childrearing); and display in bookstores labeled as "essay corners" (*essee kōnā*), shelves or areas dedicated to the genre.[11]

While some fans of Nonami's essay manga are male, she believes that male readers generally prefer sensational stories to those about everyday life. She believes that female readers are more likely to empathize (*kyōkansuru*) with characters in these run-of-the-mill stories. She has heard female readers say things like "I can totally relate to that" or "Wow, I want to do this too!" Arguably, her view that more women read essay manga might be of gender bias; however, given the genesis of essay manga by *shōjo* manga artists, as described in chapter "Essay Manga", her comment appears to hit the nail on the head. Furthermore, her description of essay manga traits is also consistent with the literature review presented in that chapter.

Nonami identifies herself as a *tōjisha* with the lived experience of Cassandra syndrome. While simultaneously dealing with Akira's Asperger's syndrome and raising two children, Nonami developed hyperthyroidism, depression, and other health problems. She believes that feelings of isolation exacerbated this stress. Whether they feel lonely or not makes a big difference in dealing with the situation; it does not matter whether the person is a family member or a hired caregiver. In such cases, she said, "The family becomes separated from the outside and suffers more as a consequence." Because of the relatively invisible nature of mental disorders, someone deeply involved in the relationship with a partner with Asperger's syndrome endures an increased level of isolation; in many cases, the wife's appeal for help gets dismissed as her menopausal reaction (*kōnenkishogai no sei*) or as a sign of suffering from hysteria (Okada 2018). It is well documented in Japan that some mental illnesses such as schizophrenia tend to be kept secret within the family because the other family members feel ashamed (e.g., Kodaira and Ito 2011; Nakamura 2013; Mukaiyachi 2016). In Nonami's view, her development of Cassandra syndrome is exacerbated by the social ideal of the wife as a "self-restrained" supporter with unconditional love. Nonami claims that well-trained therapists who can evaluate which traits of Asperger's syndrome trigger Cassandra syndrome are in short supply in Japan, a point substantiated by recently published books such as Okada (2018) and Miyao (2018).

I asked Nonami to elaborate on why she believes that the emotional burden is higher for a female neurotypical partner than for a partner of another gender. She responded by saying that the gender difference is immense when it comes to the role one plays. Although she agrees that male spouses also suffer, she blames Japan's "role-dividing system" (*yakuwari buntan mitaina mono*) in which the husband is expected to leave home for work and the wife is to take care of household chores, even if she also works. She explains that one can imagine a situation in which the husband comes home and sees his wife overwhelmed with chores piled up, their baby crying, and their home unkept; although he may feel troubled by the sight, he may still tell his wife, "Those are your responsibilities" (*sore wa kimi no shigoto dakara*). There is also an unspoken rule that parenting is the mother's domain, which means a larger portion of chores for the wife with children than for her husband, whether she works in the home or outside of it. By contrast, the husband is permitted to "forget about home," once he is at the workplace. By going to work and getting fresh air outside the next day, he can forget about the chaos at home. Therefore, male spouses have a much easier time in switching their mental gears, in Nonami's view. Working at home as a manga artist, she did not have such a privilege. Her view seems to be supported by one of Brinton's (2017) findings that older Japanese couples tend to maintain a stronger division of gender-based roles in handling housework than younger couples (p. 148).

Nonami said in-laws and acquaintances tend to defend the Aspie by saying things such as "He earns bread and butter for the family. Don't make a big deal out of things like that (*sore gurai wa amaku minasaiyo*)" or "It's typical of men" (*otokono hito wa minna sō nan dakara*). Indeed, in one of her manga episodes, Tsuna becomes more confused and depressed when one of her friends says, "Men are all like that" (*dansee wa minna sonna monoyo*) (volume 5, p. 56). In our interview, Nonami claimed that to run the household smoothly, wives have to carry everything on their shoulders (*zenbu showanakya ikenai*) and that she herself subscribed to the convention that the wife's responsibility is to take care of the home. "That belief is definitely shared by the majority (of married women)," she said, asserting that it is the wife who gets scolded when her husband looks sloppy. She also pointed out that when the wife is an Aspie, her mother feels guilty; although her husband's family may see her as a "useless bride" (*damena yome*), they tend to express no more criticism beyond that. In her view, the neurotypical husband is rarely criticized by these folks for not doing enough to support his AS wife. She emphasized that a deep-rooted gender difference exists in Japanese culture, considering how differently in-laws and acquaintances react if the neurotypical spouse is female in

Japan. From these comments, it is clear that Nonami believes that general society holds the wife of an Aspie at a higher level of accountability. Quoting Dr. Miyao, Nonami stated that this condition is not limited to domestic partners; it may be experienced by anyone with a close relationship to the patient, such as parents, children, siblings, and coworkers. The condition results from communication breakdown, especially from the inability to engage in bilateral communication. Communication with an Aspie tends to lack a transactional type of interaction in which the person receives and smoothly decodes the intention of the other person. Yet not everyone in such a relationship develops Cassandra syndrome. How the person feels connected to her partner also influences whether she develops the condition or not.

When Nonami became ill with depression and other serious health issues, she sometimes could not get out of bed. One such day, she reached a turning point. While in bed, she noticed that Akira was letting their children play video games in the next room instead of helping them with their schoolwork. She thought to herself, "I can't die! If I die, these children will be left for him to raise." Sensing a strong need to live for the sake of their children, she told herself to get some help to become well. She emphasized that women—daughters and wives, in particular—are more likely to be sacrificed by family needs. Nonami thinks this tradition is still alive. The problem is, she said, that the idea is so deeply ingrained in many women that they themselves take it for granted and become hypercritical of women who refuse to obey that tradition in Japan. Even though the social ideal of *ryōsai kenbo* (the good wife and wise mother) widely discussed in the Meiji period (1868–1912) may be obsolete nowadays, Japanese people like Nonami feel that Japan is still a country where individuals' goals and choices are deeply connected with their family responsibilities. Unlike *sengyō shufu* (so-called professional housewives) whom Vogel (2016) examined from 1958 to 1960, wives in twenty-first-century Japan, especially those with their own careers, are more likely to "embrace a sense of independence in their marital relationships" (Fujita 2016: 122). However, not prioritizing such responsibilities over one's own needs may still be perceived as an act of selfishness, especially in caring for the ill in the family. While we need more research to truly understand how widely opinions like Nonami's are held among Japanese wives, it should be reiterated that Nonami's story is the experience and reflection of one neurotypical spouse of an Aspie.

Nonami's Drawing Styles and Storytelling Characteristics

This section presents how the mostly invisible nature of Akira's or Tsuna's mental health struggles is represented through a variety of visual and textual strategies and how this manga artist came to decide on such artistic representations.

Artistic Characteristics

In creating the *Akira-san* series, Nonami used the common characteristics of essay manga described in chapter "Essay Manga". To draw the main characters, in particular, Nonami adapted the "cute design of Japan's popular culture characters like Hello Kitty," in her words. Her characters are typically drawn in the manga style of *san-tōshin* ("one-third is head") in their head-body ratio—sometimes the ratio is close to *ni-tōshin* ("two-thirds") in her character design. Amano added that Nonami draws characters even shorter in height and with more round lines than the average essay manga drawings. Figure 6.2 shows how Nonami employed a simplistic and cute drawing style for all characters presented against an unpretentious background made out of varying screentone patterns. That was intentional as she wanted to make the *Akira-san* series fully reader-friendly. Because the public was not familiar with the term *Asperger's syndrome*, she estimated that most readers would not understand this medical condition easily. Thus, to reduce the reader's cognitive load, she steered clear of visually complicated images. She also consciously selected the adorable, *san-tōshin* design (see Akira's head-torso ratio in bottom panel in Fig. 6.2) to lighten dark topics, such as Akira's debt problem and Tsuna's depression and other health issues resulting from troubles at home. Furthermore, Nonami thought people would not read her manga if the characters looked too dreary. She said, if a potential buyer thought, "Oh, it looks cute," they would be more likely to purchase it. All the characters that appear in the *Akira-san* series fit into a lineage of "healing-type" (*iyashi-kei*) characters or "wobbly characters" (*yurukyara*), like Pikachu in the Pokémon franchise and Gudetama, a Sanrio animation character, whose aesthetics make them more approachable and emotionally comfortable (Occhi 2012, 2018; Tan 2014).

Nonami drew Tsuna with big, round eyes and semi-short hair, expressing various emotions (Fig. 6.3). For example, Tsuna is depicted as a vibrant woman with heightened energy (implied with short motion lines, or *dōsen*, beaming out of her head) prior to developing Cassandra syndrome. In

Fig. 6.2 Image by Nonami Tsuna. (Courtesy of Kosumikku Shuppan)

Fig. 6.3 Image by Nonami Tsuna. (Courtesy of Kosumikku Shuppan)

contrast, to capture Akira's real-life personality, she made sure that the character does not show much emotion on his face, nor does he express his feelings in words. For example, in Volumes 1 and 2, Akira is mostly drawn with eyes that look like black dots, especially when he "freezes" in panic and becomes speechless (Fig. 6.4). She figured that the best way to depict his AS-related characteristics is by showing them visually to the reader. By drawing Akira's face as expressionless, she succeeded in creating a symbolic representation of his mental condition, which contrasts with Tsuna's vivid facial expressions.

Amano said when she first met Akira, she was so struck by how well he resembled Nonami's illustration that she almost burst into laughter. Skilled manga artists are good at creating characters by condensing detailed characteristics into a symbolic image. As Nonami reiterated, "Since this is manga, I intentionally avoid creating a realistic image of a person." Nonami told me that she extracted that image from many of their family or souvenir photos, which portray Akira always with the same facial expression and posture, facing the camera directly. She said laughing, "That's really astonishing. He never tries to fit an occasion by posing differently. So, he can be easily identified among all the other parents in (our children's) school photos. It's as if he has only one fixed stance to display."

Nonami also uses speech bubbles masterfully to represent the contrast between the Aspie husband and his neurotypical wife. For example, for Akira, she uses one plain form constantly to imply his relatively emotion-free speech mannerism. In reality, Akira maintains a steady volume while speaking and hardly ever speaks loudly. For Tsuna, she employs various shapes of speech balloons to illustrate her emotional ups and downs: a fluffy balloon to express her happy tone, and a balloon with jagged edges to illustrate her shouting, for instance (Fig. 6.5). She sometimes displays Akira and Tsuna's single speech bubbles side by side, within one frame, to further emphasize the contrast between the characters.

Although the exact symptoms and degree of intensity of Asperger's syndrome differ from individual to individual, Nonami illustrates in numerous episodes the three major areas of difficulties commonly associated with Akira's condition: social skills, communication issues, and trouble relating to others' feelings (to be discussed as "three pillars of Asperger-associated conditions" in more detail in the last section). To convey the difficulties of communicating with someone with Asperger's syndrome, she drew how Akira's "freeze" mode sets in suddenly (see middle panel in Fig. 6.5), which frustrates Tsuna immensely because she needs to solve serious household problems with his input and cooperation.

Fig. 6.4 Image by Nonami Tsuna. (Courtesy of Kosumikku Shuppan)

Fig. 6.5 Image by Nonami Tsuna. (Courtesy of Kosumikku Shuppan)

Specific Drawing Techniques

Manga artists utilize various techniques to express the character's inner thoughts. Yet, visually depicting the invisible symptoms of mental health issues would be a challenge to manga artists. I asked Nonami to elaborate on her choice of techniques such as framing, narration, speech balloons, and other strategies to represent those emotional nuances. As stated earlier, she uses some standard manga signs such as motion lines (*dōsen*) to indicate the character's bountiful energy (e.g., top panel in Fig. 6.3). To visually indicate Akira's state of being in shock or utterly confused, she makes the character stand bold-upright (*bōdachi*) and appear "petrified" (e.g., middle panel in Fig. 6.5).

In contrast to emotionless Akira, Tsuna is frequently drawn in various poses and movements indicating how active and lively she is, corresponding to the state of her mental health. For example, her emotional state is dramatically illustrated with the use of a screentone shade, an *onyu* (sound effect) "*Gaan*" to signify her shock, and her "blank white eyes," a typical manga sign used to imply a feeling of shock in the bottom panel (Fig. 6.6). The scene depicts the moment in which Tsuna receives the upsetting news from her editor that her workload has been reduced (thus, smaller income) in the midst of a financial crisis caused by Akira's unemployment and sizable debt.

Nonami told me that she had experimented with different artistic strategies to make invisible matters more discernible to the reader. She confessed, however, that she always has a hard time figuring out the best way to portray things comprehensibly. One such example is how to convey to the reader her feeling that "something is out of place" (*iwakan*) about her husband. One trick she used is to display what Nonami felt at that time as Tsuna's inner thoughts (e.g., "What? What is the matter with him?") through the thought bubbles (e.g., middle panel of Fig. 6.6). Because the reader tends to skip over a simple narration, Nonami accentuates the scene with Tsuna's impactful facial expressions (with "*Gaan*") while the character narrates.

Another strategy is to tell her inner state metaphorically with her thoughts as shown in Fig. 6.7 where Tsuna's pale-looking figure fights to cut herself from a tentacle strapped around her ankle, which represents each of her guilt feelings textually expressed in a boxed field (e.g., "guilt about not being able to notice Akira's Asperger's for a long time" in the first panel). Comically, she adds making Akira's story into a comic as

Fig. 6.6 Image by Nonami Tsuna. (Courtesy of Kosumikku Shuppan)

Fig. 6.7 Image by Nonami Tsuna. (Courtesy of Kosumikku Shuppan)

another source of guilt but allows Tsuna to release from that guilt with a handwritten comment right below which says, "But he gave me the permission!" (see bottom left panel). In sum, whatever artistic strategy she uses, she prioritizes readability.

Readability Considerations

Nonami believes that essay manga is the genre best suited to deliver her story in an approachable way. Although the genre capitalizes on simplistic drawings overall, essay manga tends to include a character's long monologue or too much technical information within in one panel. Nonami consciously avoids too much use of text and tries to keep each frame simple for readability. She also separates the main story of manga from Dr. Miyao's comments by inserting his input as a one-page medical commentary at the end of each episode (e.g., "Dr. Miyao's description of AS" in Volume 1). Another consideration that she has for the *Akira-san* series is keeping medical information about mental health problems to a minimum, especially if the reader does not have a strong background knowledge of psychology. Nonami also said, "I keep each episode relatively short because it's hard for the reader to follow a long story with a complicated plot on such a serious topic like Asperger's." She knew that manga readers do not welcome a story with numerous panels packed into one page. Furthermore, she keeps sentences relatively short, even when conveying dense information, as shown in Figs. 6.2–6.7. "A good bathroom book" (*toire tokani oite, chotto patto hiraita tokoro kara demo yomeruyouna*) is what Nonami conceptualizes as an essay manga of high readability. In other words, it is the kind of reading one can stop at any point to leave the bathroom and then pick right back up the next time around. For that end, Nonami keeps each episode fairly short while making sure that each short episode is part of a larger storyline. "Even though the episode itself carries a lot of information, it's easy to digest the story if you can read quickly from the first to the last page," she said. With these considerations, Nonami has created each volume as a collection of short episodes, rife with humor and educational information about Asperger's and Cassandra syndrome.

Amano added that one of manga's strengths is that the medium easily disseminates technical knowledge such as medical information widely to the public. However, as a manga artist, Nonami is acutely aware that reading itself can be a hurdle, as comprehending text can be a cognitive load heavier than watching or listening a story. She uttered with a giggle, "I'm hoping that, as they (readers) are flipping through my manga's pages, they spot an exciting scene and decide to buy it!" Both Nonami and Amano pointed out that reading manga is "a volitional act"; that is, consumers make their own choice to pick up and read that particular manga. By contrast, consumers can just sit there to watch and listen to the input in film and television dramas; the strength of these media is that the viewer can remain relatively passive. Nonami said, "Even if the viewer gets bored in the middle, they can sit still until the story becomes interesting. But, for manga, if the reader finds the story boring, they will stop turning pages and put the book down. That's the end of it. That's how manga is different from the other media." Therefore, to attract consumers' attention and help them read the story all the way to the end, the artist must utilize all sorts of gimmicks. She believes that what separates manga from the other forms of popular media is that it has an infinite number of approaches to artistic expressions (*manga wa hyogen ga mugen*). "For example, you can adopt a novel-like expression or a film-like panel structure in drawing manga," she added. In her opinion, this infinity of expressional means is manga's greatest strength.

Reception and Generalizability

How did Nonami's *Akira-san* series impact on public awareness of Asperger's or Cassandra syndrome in Japan? Nonami told me that years ago, when she told people her story, she usually had to explain Akira's neurological disability at the beginning. Before 2011, the year in which Nonami published Volume 1 of the *Akira-san* series, public awareness of Asperger's syndrome was quite limited. That is why Nonami's manga made a big splash and received more than one hundred messages from her readers immediately after the publication of that volume (Sankei Newspaper News 2017). But her "breakthrough" year was 2014, when the fourth volume was published in 2014, her manga came to national attention: the series was featured on "Morning Number One" (*Asa-Ichi*), a widely watched talk show broadcast by NHK (*Nippon Hōsō Kyōkai*, Japan's public broadcasting station, sponsored by the Japanese government), when they ran a series of discussions about Cassandra syndrome. Thanks to this broadcast, her *Akira-san* series became explosively popular (*ninkido ga bakuhatsu*), and 22,000–23,000 copies were sold that year. Although she humbly pointed out that the number was not extremely large ("Especially because essay manga tends to reach a small market," she added), it is clear that her work made a significant impact on the public understanding of both Asperger's and Cassandra syndrome.

Nonami pointed out that Hosokawa Tenten's *My Significant Other Has Got Depression* (2006) was the first manga written from a wife's point of view about what it was like to live with a husband suffering from depression. However, when it comes to Asperger's syndrome, she believes that her *Akira-san* series was the pioneering work written from a wife's point of view. The *Akira-san* series has been cited as a pioneer work in blogs and publications about Asperger's syndrome (e.g., Shūkan Asahi 2014; *Quarterly B!* 2016; Pixiv Essay 2016; Watanabe 2020). Following its publication, more manga titles have been published on this topic, including *Gaki no tameiki* (2011–2013), a title analyzed in chapter "Okita Bakka's *Gaki-Tame* Series (2011–2013): A Memoir of a "Troublemaker" Aspie Girl" of this book, and the two-volume series *My Girlfriend has Developmental Disorders* (*Boku no kanojo wa hattatsu shōgai*) published in 2013 and 2015 respectively.[12]

Amano pointed out in the interview that *ikizurasa* (difficulty in living) is a concern frequently expressed by many Aspies in Japan. In fact, *ikizurasa* is a frequently expressed theme in Asperger's syndrome memoirs (e.g., Ayaya 2013; Komichi 2009; Niki 2010). Although Nonami's series does not directly address the concept of *ikizurasa*, the manga illustrates many incidents about how Akira runs into problems with other people in three areas of difficulties associated with Asperger's syndrome: communication problems, social relationship problems, and social imagination difficulties (e.g., difficulty to see things from someone else's perspective). These areas are exactly what Lorna Wing delineated as three pillars of Asperger-associated conditions (Iwanami 2017). Similarly, the series depicts many episodes in which Akira and Tsuna's communication tends to be void of *a-un no kokyū* (*a-un* breathing), a Japanese phrase implying a high level of harmonious communication among like-minded people.[13]

Japan's provision of support for people with Asperger's syndrome is still limited (Hirose 2015; Mori 2018; Okuyama 2020). As Amano pointed out in our interview, Aspies can receive a *shōgai techō* (a disability passbook issued as a disability certificate by the local government), depending on the degree of severity of their condition (e.g., Bureau of Social Welfare and Public Health, Tokyo Metropolitan Government n.d.). To receive the passport, one must meet three requirements: (1) needs as an AS *tōjisha*, (2) evidence that indicates his/her inability to demonstrate sociability (*shakaisei ga itonamenai*), and (3) a doctor's diagnosis. However, Akira's Asperger's does not cause depression, nor does it lead to any other *niji shōgai* (secondary disabilities); he is able to work and is currently employed. Therefore, even though he was officially diagnosed with Asperger's, his condition is not considered a clinically serious disability. Thus, he is not qualified to receive the disability certificate. Nonami said that what is clear to her is that Akira himself does not feel negatively affected by his condition.

Nonami told me that what affected her family the most was Akira's accumulated debt and unemployment. She said, "That was really unbelievable (*futsū no hito dewa kangae rarenai*)." In her view, if it had been either debt or job resignation, but not both simultaneously, the situation might have been more manageable. In Nonami's words, it is Akira's Asperger's condition that makes him lack common sense (*shakaisei ga ichijirushiku ochiteiru*). Interestingly, Hirose (2015) points out that the reduced capacity for imagination common among Aspies (the third type of disability delineated by Wing's studies) seems to debilitate their ability to envision the future and infer appropriate financial planning. Although similar cases are reported anecdotally (e.g., Kurage and Terashima 2015; Kuon 2016; *Shūkan Asahi* 2014), it is important to point out that not everyone with this condition ends up with serious financial trouble as Akira did.

To Nonami herself, the *Akira-san* series is most noteworthy for its contribution to empower women, especially mothers and wives of Aspies, who may be at risk of developing Cassandra syndrome. This is because whether the person is informed of the risk or not makes a huge difference in the degree of her suffering. The knowledge is powerful, as it helps explain to the afflicted, as well as to the general public, what she is going through, and it helps raise awareness of the condition in the local community. Publications such as Miyao's (2018) books about Cassandra syndrome are also important sources of information. For example, they have helped in children's *ryōiku* (rehabilitation of autistic children, mentioned in chapter "Okita Bakka's *Gaki-Tame* Series (2011–2013): A Memoir of a "Troublemaker" Aspie Girl"), especially those who suffer from interpersonal communication problems. Nonami proudly called the *Akira-san* series "the first manga story about a wife with an adult *hattatsu shōgai haigūsha* (spouses of individuals with *hattatsu shōgai*)."[14] These days, Nonami gives public talks about her Cassandra experience approximately four times a year. Although talk requests usually come from wives married to Aspies, she has begun to receive requests from counselors who want to become familiar with Cassandra syndrome for their professional knowledge. Her most recent talk was given at a seminar organized by a company that hires people with disabilities. She feels that her words extend beyond the small circle of wives and counselors and that more people are taking an interest in Cassandra syndrome.

Efforts to make the public aware of Cassandra syndrome are finally bearing fruit. Fortunately, self-help groups (*jijo gurūpu*) have sprung up in Japan (Shūkan Asahi 2014). One self-help group called "Khan's wives" (*Hān no tsumatachi*) in Tokyo was founded in 2013 by a middle-aged woman, Sora, and garnered approximately 200 members within a year (Shūkan Asahi 2014, 125). Similar to Nonami, Sora was married to an Aspie who was unable to manage money and experienced Cassandra syndrome. Her blog is dubbed "A Shelter for Cassandras" (*Kasandora kakekomidera*) (http://blog.livedoor.jp/khan2013214/). In 2016, *Quarterly "B"! (Kikan Bi!)*, a popular magazine about mental health issues, featured Cassandra syndrome in three consecutive issues (no. 123, 124, and 125). The first issue describes Nonami's manga and introduces the reader to several peer support groups located in major Japanese cities (Quarterly "B"! 2016). Books and academic articles have also been written about the syndrome (e.g., Edogawa 2017; Okada 2018). A recently published manga about a "Cassandra wife"—Ruon's (2020) *A Cassandra Wife's Humorous Battle* (*Kasandora tsuma no komikaru funtōki*)—is specifically geared toward readers who endure the stress of living with their Aspie husbands during the COVID-19 pandemic. Although it is my speculation, this comic and recent non-comic books such as Saijō Sarayo's (2015) *A Cassandra Wife's Memoir* (*Kasandora tsuma no taikenki*) and Shingyo Yuiko's (2020) *Does My Husband Have Autism Spectrum Disorder?* (*Watashi no ottowa hattatsu shōgai?*) are very likely inspired by the popularity of the *Akira-san* series. Because of these publications, people in Japan now are more familiar with the term Cassandra syndrome compared to a decade ago.

Nonami is aware that her experience is not generalizable to every neurotypical and neuro-atypical partnership. Still, as a tōjisha, she wants to continue representing the voice of many other "Cassandra" wives by using her visual art skills to fight Japan's traditional idea that "a wife should emotionally support her husband" (*tsuma nandakara otto wo sasae nakya*). She strongly believes that the wives who are bound by this old-fashioned idea need to know that they should be free from that burden (*sonna noni shibararenakute iindayo*), to understand that they must prioritize what is important to them (*jibun no ikitaikoto wo mein ni kangaeru*), and first and foremost to think what makes them happy (*jibun ga dō ikitara shiawaseka*). Although she acknowledges that such patriarchal ideas are also ingrained in other societies and that women have found diverse ways to negotiate gender norms for women in the family, she still stresses her main message for any reader: "You should not devote your whole self to your family alone. Take good care of yourself first (*mazu jibun wo taisetsu ni*)."

Nonami's current ambition is a wider dissemination of her "Cassandra" wife story through television broadcasting, since she is acutely aware of its advertising power. She said, "To tell the truth, I want my manga to be adapted into a television drama. I think that would help my story reach out to those who don't customarily read manga." Indeed, Hosokawa Tenten's essay manga, "*Tsure ga utsu ni narimashite*" (*My SO Has Got Depression* 2009), which Nonami mentioned earlier in the interview, was first adapted into an NHK television drama series in 2009 and then made into a film and released by Toei in 2011. That autobiographical comic features Hosokawa's marriage with an employee of a demanding IT company who develops major depression. As if it were a chain reaction, Hosokawa's hit manga inspired another *mangaka*, Tanaka Keiichi, to publish in 2017 his own depression memoir, *Utunuke*, which has also been made into a television show recently. The next two chapters analyze these two popular graphic memoirs.

Notes

1. Volumes 1 and 2 were published in 2011, Volume 3 in 2012, Volumes 4 and 5 in 2014, Volume 6 in 2015, Volume 7 in 2016, and Volume 8 in 2017.
2. Iwanami (2017) reports the ratio of Asperger's syndrome as five out of 1000 people and points out that there are more men than women afflicted with this neurological disorder. This gender disparity partially accounts for more "Cassandra" cases of wives than husbands. As a psychiatrist himself, Iwanami observes from clinical cases that highly intelligent men with autism spectrum disorder may have a relatively trouble-free childhood and adolescence, especially if they grow up in a supportive environment, and are likely to have a delayed diagnosis of the disorder.
3. According to *Sankei News* (2017), Akira's debt was reported to be three million yen, equivalent to approximately $30,000. According to a family history presented in Volume 4, Tsuna finds out about the debt before

Akira is diagnosed with Asperger's syndrome. The episode of debt discovery is frequently depicted as a traumatic moment for Tsuna throughout the series.

4. Because English-language articles refer to a person with Asperger's syndrome as an "Aspie" and people who do not have that condition as "neurotypical" individuals, both expressions (e.g., a wife who is not an Aspie is identified as "a neurotypical wife") are used in this chapter. I also analyzed some manga on the topics of Autism Spectrum Disorder (ASD) in *Reframing Disability in Manga* (2020).
5. At this point, Nonami began to work with Amano as the main editor of the Akira-san series along with a collaborative editor, Nishida Sachie. Starting with Volume 6, Amano became the sole editor of the series.
6. Although receiving favorable responses from readers and being featured on NHK's Morning Number One, Nonami has also received criticism about her depiction of Akira in the *Akira-san* series. For example, some customer reviews at Amazon.com accused her of depicting Asperger's syndrome with hostility and her manga as being merely self-pity about how tough her life has been because of her husband's condition. She believes that a fraction of Aspies felt they were victimized by her manga, asserting that it was "putting us in the wrong." In the interview, although she defended herself by saying, "If you read [my series] carefully, you wouldn't have come up with such a conclusion anyway," Nonami now knows that AS tōjisha also read the series even though it targets neurotypical readers and is more careful not to hurt their feelings.
7. As mentioned in this chapter, both Dr. Miyao, who diagnosed Akira with Asperger's syndrome, and the counselor Takiguchi helped Nonami during her development of the Akira-san series. Together, Miyao and Takiguchi translated Maxine Aston's *Asperger in Love* (2003) into Japanese.
8. Although I am certain it is her penname (because she asked me if I wanted her to sign with her penname or real name on the interview permission form), I did not ask her how she came to use it. Unlike some of my other interviewees with unusual names (e.g., Okita Bakka, Hosokawa Tenten), Nonami Tsuna is an ordinary Japanese name, and I did not see any importance in inquiring about the origin of her penname.
9. Japan's comics encompass different genres including *seinen* manga, *rediisu komikkusu* (ladies comics), *gag* manga, and essay manga (also called *taiken* comics). *Seinen* manga is the genre marketed toward young adult male readers with representative manga titles being *Ghost in the Shell* (*Kōkaku Kidōtai*) created by Masamune Shirow and *The Gourmet* (*Oishinbō* 1984), a manga about cooking written by Kariya Tetsu and illustrated by Hanasaki Akira. In the 1980s, the genre of ladies comics, also called *josei* manga, was developed for female manga readers older than those of *shōjo* manga (comics aimed at youth readers).

10. Essay manga is a genre developed in the 1980s in the Japanese manga industry and is also called ***komikku essei*** (comic essay) and ***taiken essei*** (experiential essay) interchangeably; during the interview, Nonami herself switched between ***komikku essei*** and ***taiken essei*** in reference to her genre. Although, other labels such as ***rupo manga*** (reportage manga, especially for its investigative content), ***watakushi manga*** (I-manga), and *jitsuyō* (how-to) manga exist for minor use, most frequently and interchangeably used labels are ***essay manga*** and ***comic essay*** (Kawahama 2007). *Nichijō* (literally "everyday") manga is the genre of comics that tell slice-of-life stories. Similarly, *taiken* (experience) manga is the genre of comics that depict authors' own lives and experiences. In our interview, Nonami used both *taiken* manga and essay comics interchangeably. In manga parlance, as well as in social media and advertisements nowadays, the label of essay comics (or essay manga) is prolific; therefore, I use the term ***essay comics*** hereafter in this article.
11. In our interview, Amano suggested that a related label is "*josei essei komikku*" (women's essay comics) and that one of its popular themes is *Tabi ni ikō* ("Let's go traveling") such as Takagi Naoko's *Hitoritabi ichinensee* (*Solo Traveling Freshman*), a manga title published by KADOKAWA in 2006.
12. As stated in that chapter, Okita Bakka, author of the *Gaki no tameiki* (2011) series, was inspired to write her graphic memoirs by Nonami's *Akira-san* series.
13. The word *a-un* refers to a pair of syllables, figuratively representing the beginning and ending of life, a set of inseparable truths, in Japanese Buddhist mythology (Okuyama 2015).
14. It is important to clarify that her use of *hattatsu shōgai* (developmental disorders) in this context is misleading because, as explained in the previous chapter, *hattatsu shōgai* is a label for a condition comprised of Asperger's syndrome and other disabilities. Akira has no learning disability and graduated from an elite university, nor does he have ADHD. However, *hattatsu shōgai* is often used interchangeably with Asperger's syndrome as in this case.

References

Aston, Maxine. 2001. *The Other Half of Asperger Syndrome*. London: The National Autistic Society.

———. 2003. *Asperger in Love: Couple Relationships and Family Affairs*. London: Jessica Kingsley Publishers.

———. 2009/2020. *The Asperger Couple's Workbook*. London: Jessica Kingsley Publishers.

Ayaya, Satsuki. 2013. "Tōjisha Studies and Self-awareness (*Tōjisha kenkyū to jiko-kan*)." In *A Study of Tōjisha Studies* (*Tōjisha kenkyū no kenkyū*) edited by Ishihara Kouji. 177–216. Tokyo: Igakushoten.

Brinton, Mary C. 2017. "Happiness at work? Marital happiness among Japanese housewives and employed wives." In *Life Course, Happiness and Well-being in Japan*, edited by Barbara Holthus and Wolfram Manzenreiter, pp. 138–157.

Bureau of Social Welfare and Public Health, Tokyo Metropolitan Government. n.d. (*Tōkyōto jidō sōdan sentā/jidō sōdanjo, tōkyōto fukushi hokenkyoku*) at https://www.fukushihoken.metro.tokyo.lg.jp/jicen/ji_annai/a_techou.html (retrieved October 2, 2020).

Edogawa, Rie. 2017. *Kasandora shōkōgun deshita* (I had Cassandra Syndrome). Tokyo: Bungeisha.

Goffman, Erving. 1963. *Stigma: Notes on the Management of Spoiled Identity*. Englewood Cliffs, NJ: Prentice-Hall, Inc.

Fujita, Noriko. 2016. "*Tenkin*: New Martial Relationships and Women's Challenges in Employment and Family." *U.S.-Japan Women's Journal*, no. 50: 115–135.

Hirose, Hiroyuki. 2015. *A Book to Read When You Suspect "Asperger's"* (*Moshikashite Asuperugā to omottara yomu hon*). Tokyo: Nagaoka Shoten.

Hosokawa, Tenten. 2006. *My Significant Other Has Got Depression* (*Tsue ga utsu ni narimashite*). Tokyo: Gentosha.

———. 2009. *After My Significant Other Has Got Depression* (*Sonogo no tsue ga utsu ni narimashite*). Tokyo: Gentosha.

Iwanami, Akira. 2017. *Hattatsu shōgai* (*Developmental disabilities*). Tokyo: Bunshuu-shinsho.

Jegatheesan, Brinda. 2009. "Cross-cultural issues in parent-professional interactions: A qualitative study of perceptions of Asian American mothers of children with developmental disabilities." *Research & Practices for Persons with Severe Disabilities*, 34, 123–136.

Kariya, Tetsu, and Akira Hanasaki. 1984. *The Gourmet* (*Oishinbō*). *Volume 1*. Tokyo: Shogakukan.

Kawahama, Hideaki. 2007. "Essay Manga's Process of Establishment—Representations of Manga Artists Themselves in Special Issues of Shōjo Comics [*Essei manga no seiritsu katei—bessatsu shōjo comics ni okeru mangaka no hyōsō ni chakumokushite*]." *Manga Kenkyū*, No.12, 62–70.

Kayama, Misa, and Wendy Haight. 2014. *Disability, Culture, and Development: A Case Study of Japanese Children at School*. New York: Oxford University.

Kodaira, Tomoe, and Takehiko Ito. 2011. "When mental health patients are the primary actors [*'Tōjisha ga shujinko' no jissen no arikata wo kangaeru: tōgōshicchō tōjisha ni yoru naratibu wo tegakari ni*]." *Wako University Sōgōbunka Kenkyūjo Nenpō "Tōzai-nanboku"*: 130–140.

Komichi, Moko. 2009. *Atashi kenkyū: Jiheishō supekutoramu* (*I-research: Autism Spectrum*). Tokyo: Kurieitsu Kamogawa.

Kuon, Minami. 2016. "Cassandra syndrome: Peer advice" (*Kasandora shokogun: chie wo kyoyusuru nakamatachi*). *Quarterly B! (Kikan Bi!)* volume 124, September 2016, 34–39.

Kurage and Terashima, Hiro. 2013. *My Girlfriend has Developmental Disorder* (*Boku no kanojo wa hattatsu shōgai*). *Volume 1.* Tokyo: Gakken Purasu.

———. 2015. *My Girlfriend has Developmental Disorder* (*Boku no kanojo wa hattatsu shōgai*). *Volume 2.* Tokyo: Gakken Purasu.

Masamune, Shirow. 1991. *Ghost in the Shell* (*Kōkaku Kidōtai*). Kodansha.

Miyao, Masutomo. 2018. *A Book to Solve Cassandra Mothers' Distress* (*Kasandora no okāsan no nayami wo kaiketusuru hon: Hattasu shōgai no otto ni furimawasarenai tameni*). Tokyo: Kawadeshobo-shinsha.

Mori, Rafushia. 2018. "Days with My Asperger Husband." (*Asuperugā no otto tono hibi*). *We*, June/July issue, no. 215: 73–75.

Mukaiyachi, Ikuyoshi. 2016. *Mental Disability and Church* (*Seishinshōgai to kyōkai*). Tokyo: Inochinokotobasha.

Nakamura, Karen. 2013. *A Disability of the Soul: An Ethnography of Schizophrenia and Mental Illness in Contemporary Japan.* Ithaca, NY: Cornel University Press.

Niki, Rinko. 2010. Escape from the epicenter of vulnerability (*Naetsubo no jiraigen kara tonari e nogarete*). In *Beckoning freaks: Disability studies of culture and representation* (*Temaneku furīku: Bunka to hyōgen no shōgaigaku*), edited by Kuramoto Tomoaki, 249–277. Tokyo: Seikatsushoin.

Nonami, Tsuna. 2011. *My Husband, Akira, Has Asperger's Syndrome* (*Akira-san wa asuperugā*) Volume One. Tokyo: Kosumikku.

———. 2014. *My Husband, Akira, Has Asperger's Syndrome Volume Five: Husband Akira is an Aspie, Wife is a Cassandra* (*Akira-san wa asuperugā, okusan wa kasandora*). Tokyo: Kosumikku.

———. 2015. *My Husband, Akira, Has Asperger's Syndrome Volume Six: Asperger's and Cassandra* (*Asuperugā to kasandora*). Tokyo: Kosumikku.

———. 2016a. *My Husband, Akira, Has Asperger's Syndrome Volume Seven: Finding Hints of Happiness* (*Mitsukeyo egao no hinto*). Tokyo: Kosumikku.

———. 2016b. *Uchi no komatta-san* (*Our Mischief-Maker*). Tokyo: Riido Publisher.

———. 2017. *My Husband, Akira, Has Asperger's Syndrome Volume 8: A Marriage without the Knowledge of Asperger's (Akirasan wa asuperugā, Asuperugā to shiranaide kekkonshitara tondemonaikoto ni narimashita*). Tokyo: Kosumikku.

Nonami, Tsuna and Yagishita, Noriko. 2013. *Conversation Techniques for People with Asperger's Syndrome* (*Hattatsu shōgai ga aruhito no tamen o mirumiru kaiwaryoku ga tsuku nōto*). Tokyo: Kodansha.

Occhi, Debra J. 2012. "Wobbly aesthetics, performance, and message: Comparing Japanese *kyara* with their anthropomorphic forebears." *Asian Ethnology*, 71(1): 109–132.

———. 2018. "Kumamon: Japan's surprisingly cheeky mascot." In *Introducing Japanese Popular Culture*, edited by Alisa Freedman and Toby Slade, pp. 13–23. New York: Routledge.

Okada, Takashi. 2018. *Cassandra Syndrome: If Someone You Know May Have Cassandra* (*Kasandora shōkōgun: Mijikana hito ga asperuga dattara*). Tokyo: Kadokawa.

Okita, Bakka. 2011 [2012, 2013]. *A Kid's Sighs* (*Gaki no tameiki*). Volumes 1–3 Tokyo: Kodansha.

Okuyama, Yoshiko. 2015. *Japanese Mythology in Film: A Semiotic Approach to Reading Japanese Film and Anime*. New York: Lexington Books.

———. 2020. *Reframing Disability in Manga*. Honolulu: University of Hawai'i Press.

Pixiv Essay. 2016. Why is Essay Manga Popular in the Social Media? New Business Tactics of Publishers (*Essei manga ga SNS de ninki no riyu towa? Shuppan-gyokai mo aratana ugoki*). 2016, June 13. https://ddnavi.com/news/305909/a/ (retrieved October 2, 2020).

Quarterly "B"! (*Kikan Bi!*). 2016. "How These Wives Got Out of Their Trap (*Tsuma toshite no kurushimikara do nukedashita?*)." *Quarterly "B"!*, 2016, June, no.123: 40–44.

Ruon. 2020. *A Cassandra Wife's Humorous Battle* (*Kasandora tsuma no komikaru funtōki*). Amazon Kindle.

Saijō, Sarayo. 2015. *A Cassandra Wife's Memoir: Recovery from An Emotional Scar* (*Kasandora tsuma no taikenki: kokoro karano kaifuku*). Tokyo: Seiwa Shoten.

Sankei Newspaper News. 2017. "Akira-san has Asperger's syndrome: The truth of 'normalcy' found by an artist who has been writing about her husband in manga (*Akira-san wa asuperugā: Otto wo kakitsuzukeru mangaka ga mitsuketa 'futsu nante dokonimo nai'*)." https://www.sankei.com/life/news/171209/lif1712090002-n1.html (retrieved October 2, 2020)

Shingyo, Yuiko. 2020. *Does My Husband Have Autism Spectrum Disorder?: How Cassandra Wives Find True Happiness* (*Watashi no ottowa hattatsu shōgai?: Kasandoran tsumatachi ga hontō no shiawaseo tsukamu hōhō,*) Tokyo: Subarusha.

Shūkan Asahi. 2014. "Wives Who Live with AS Husbands: Do You Know "Cassandra Syndrome"? (*'Asuperugā' no otto to kurasu tsuma tachi: 'Casandora shōkōgun' tte shitteimasuka?'*). *Shūkan Asahi* (*Weekly Asahi*). August 29, 2014: 122–125.

Takagi, Naoko. 2006. *Hitoritabi ichinensee* (*Solo Traveling Freshman*). Tokyo: KADOKAWA.

Tan, Caroline S.L. 2014. "*Gotochikyara* & *Yurukyara:* The Fusion of Pop Culture in Place Branding in Japan." *International Journal of Business and Economics*, 6(1): 75–89.

Tanaka Keiichi. 2017. *Overcoming Depression* (*Utsunuke*). Tokyo: Kadokawa.

Vogel, Suzanne Hall. 2016. *The Japanese Family in Transition: From the Professional Housewife Ideal to the Dilemmas of Choice*. [With Steven K. Vogel.] New York: Rowman & Littlefield Publishers.

Watanabe, Shohei. 2020. "'*Asuperugā' to kaizen no monogatari: Shirōto senmonka ni yoru jiheishō supekutoram shōgai no ichirei* ('Asperger' and the narrative of improvement: Lay-experts' portrayal of autism spectrum disorder in Japan)." *Ningen Shakaigaku Kenkyu Shūroku*, 15: 113–137.

Yoshida, Tomoko. 2005. *Anataga anata de arutameni: Jibun rashiku ikiru tame no asuperugā shōkogun gaido* (*To Remain Who You are: A Guide for Living in Harmony with Asperger's Syndrome*). Tokyo: Chuohoki-shuppan.

CHAPTER 7

Tanaka Keiichi's *Utsunuke* (2017): An Ode to Depression *Tōjisha*

Introduction

Tanaka Keiichi's graphic memoir, published by KADOKAWA, is an ode to those with the lived experience of depression. As a tōjisha who struggled with his illness for approximately ten years, Tanaka wrote a personalized message addressed to his fellow tōjisha in the manga *Utsunuke: Utsu tonneru wo nuketa hitotachi* (*Depression Breakthrough: People Who Got Through the Depression Tunnel*): "Dear Fellow Tōjisha, please know that there is light at the end of the depression tunnel" (p. 162).[1] This is a unique comic that is published in a multicolor format since manga is normally a monochrome medium. It consists of twenty short episodes presented in a tight space of eight pages per episode. The first three episodes depict Tanaka's own experience, followed by the episodes of sixteen other cases and an episode of Yūki Yū, chief psychiatrist of a mental clinic in Tokyo who is known for a web comic, "Understanding Mental Illnesses through Manga" (*Manga de wakaru shinryō naika*). The protagonist, "Tanaka-san," who is Tanaka himself, appears with his assistant "Kaneko" in each episode. As shown in Fig. 7.1, Tanaka-san and all the other characters are drawn to resemble the style of Tezuka Osamu. In analyzing this title, I will use the term *Utsunuke* hereafter.

Depression, or more formally, major depressive disorder, is a condition accompanying symptoms such as prolonged depressed mood and reduced interest in activities that the person previously considered pleasurable

Y. Okuyama, *Tōjisha Manga*,
https://doi.org/10.1007/978-3-031-00840-5_7

Fig. 7.1 Image by Tanaka Keiichi. (Courtesy of KADOKAWA)

(DSM-5 2013: 160). Tanaka began to experience depression symptoms as well as panic attacks in 2005. It was not until 2014, however, that his condition improved and he was able to fulfill his goal of writing *Utsunuke*. By October 2014, he had managed to keep his symptoms under control, and in December, his *Utsunuke* began to be serialized in a digital magazine, *Dejitaru Shosetsushi*. Its *takōbon* version was published in 2017. His identity as an artist and his mission-like desire to create a manga out of his painful experience helped him maintain some sense of hope.

However, *Utsunuke* was Tanaka's first manga that dealt with such a socially serious topic as a mental illness. Although it was uncharted territory to him, he was fortunately assisted by an experienced, highly capable editor who provided him with the right advice on detailed matters ranging from word choice to panel size to overall structure. This chapter details the manga's genesis, the artist's background, his drawing, and other techniques, as well as the sources of the manga's success that he articulated in our interview.

The Title and Target Readership

Utsunuke has two key phrases, *utsunuke* in the main and *utsu tonneru* in the subtitle, which contains *utsu* (short for *utsubyō*, or clinical depression). As for the first phrase, *utsunuke* literally means "coming through depression," Tanaka made up this expression to create what he calls a "hook title" (*fukku taitoru*), a title that can attract many potential readers who are curious about what it means. He explained that Kodansha, one of the major publishers, has succeeded in creating hook titles by creating titles that are either comprised of or likely to be abbreviated into four syllables. As an example, he referred to Moritaka Yūji's *Gurazeni* (*Money Diamond* 2010–2014), the title of a hugely popular baseball manga (serialized in Kodansha's *Weekly Morning* magazine).[2] "I also call it an 'enigmatic four-syllable word' (*nazo no yonmoji*) that is catchy and hooks the reader. I knew of its powerful impact, and I had been wanting to use such a title for one of my manga someday." By employing Kodansha's business tactic for "hook titles," Tanaka invented a four-syllable word *utunuke* by combining *utsu* (depression) and *nuke*, a noun derived from the verb *nukeru* (to come through). He used it in the main title to clarify what this manga is about. "I thought to myself, 'This title is easy to understand and memorize,'" he told me. According to *GQ Japan* (2019), the word *utunuke* was nominated as a popular word of the year in 2017 and has become a term of social importance to connote "recovery from depression."

The second key phrase, *utsu tonneru* (the depression tunnel), appears in its subtitle *Utsu tonneru wo nuketa hitotachi* (*People Who Got Through the Depression Tunnel*). Tanaka chose sixteen depression tōjisha who came through that tunnel, along with himself, to interview for this manga. A recurring theme of *Utsunuke*, the depression tunnel metaphorically expresses a dark space in which the afflicted are trapped alone. In the interview, Tanaka said that being in the tunnel was a "miserable and excruciating" (*tsurakute kurushii*) experience. Kaneko Akiko, his *tantō henshūsha* (manga editor), was quite skillful at discerning such impactful themes in the seventeen interviews.

As mentioned in the introduction, the primary target readership of *Utsunuke* is none other than the depressed, especially those who feel stuck in the tunnel on a journey of no end. That target was clear to Tanaka ab initio. By that time, however, several autobiographical manga, especially Hosokawa Tenten's megahit, *Tsure utsu* (covered in chapter "Hosokawa Tenten's *Tsure utsu* Series (2006–2013): A Couple's Lived Experience of Depression"), had already been published. What inspired him to write his graphic memoir? He said that it was Miyajima Kenya's book (2010) *Jibun no "utsu" wo naoshita seishinkai no hōhō* (*A Psychiatrist's Approach to the Recovery of His Own "Depression"*) about his recovery from depression. Tanaka calls it "a book of fateful encounter (*gūzen deau issatsu*)." He said to me, "Until then, I had been fearful of leading life with an incurable illness." Because of that high level of fear, when he was able to exit the tunnel, he was all the more grateful and felt compelled to do the same in return. "I had a sense of mission to bring to the world another *gūzen deau issatsu* for the many who are suffering and stuck in the tunnel," he added as he emphasized "sense of mission." In the manga's epilogue ("*Atogaki*," p. 173), Tanaka writes that he stumbled upon Miyajima's book at a convenience store one night and that it guided him to the tunnel's exit after he had wandered about inside for nearly ten years. Writing this memoir through Tanaka's own professional outlet—manga—was his way of returning Miyajima's favor. That is why he is particularly delighted when other tōjisha tell him how much his manga means to them.

However, because his primary target was depression tōjisha, he initially expected that his *Utsunuke* would never be a blockbuster like *Tsure utsu*. "It [*Tsure utsu*] did not target those who are in the midst of depression because they do not think of reading books. Depression diminishes your willpower to read anything, let alone books on how to get out of depression." Tanaka told me that *Tsure utsu* had sold well primarily because it targeted healthy readers who either have or might have a family member afflicted with depression. By contrast, he dedicated his manga to the depressed, unwilling consumers of print publications. At the same time, he also thought intuitively that unlike text-only books, they might find comics rather readable (*Money Plus* 2017). Surprisingly, many non-tōjisha readers bought his manga out of curiosity, wanting to know what would happen if depression ever hit them as it is a condition that is more widespread than previously thought. In addition, parents and children of the depressed also purchased the manga. Tanaka's choice of a catchy title and other commercially viable strategies, such as adding sixteen other cases, enabled this manga to attract a larger audience beyond a relatively smaller number of depression patients.[3] Tanaka believes another charm of the manga lies in its visualization of an invisible illness like depression (*GQ Japan* 2019). "So, I worried for nothing," he said.

Although Tanaka claims that Miyajima's book was his chief inspiration to write *Utsunuke*, it appears that his innate creative energy helped espouse this manga immensely. He told me that, while still receiving medical treatment, he was taken over by a strong desire to make his lived experience of depression into a comic story because he felt that "this hardship is a valuable experience [*kono taiken wa kichō dakara*]." When he posted his first proposal for *Utsunuke* on a social media site, he was still fighting his depression symptoms and was nowhere near the "exit" of the tunnel. "I was spurred on as a storyteller by the thought that I must make most of this abnormal experience by making it into manga," he added.

In the interview, he mentioned two other works, both written by peer manga artists, which also galvanized him into action. The first is Ueno Kentaro's (2011) manga, *Sayonara mo iwazuni* (*Without Saying a Goodbye*), which he read while he was still deeply depressed. Based on Ueno's real-life tragedy, it tells the story of a man whose wife suddenly passed away, leaving behind their young daughter. Praising Ueno's storytelling technique through which Tanaka vicariously experienced what the author went through, he told me, "I could really feel his pain as I was merging with Ueno's character in the manga." Tanaka, a gag manga artist like Ueno, was impressed by the skill used by this *mangaka* in writing about such a serious topic as death. Tanaka felt a strong kinship with this artist when reading about his desire to make use of his own tragedy for creativity in the manga's epilogue. "I saw him saying something like 'it's an illuminating [*oishii*] experience for a writer.' I sensed his desire to preserve it in the form of art." Inspired by Ueno's words, Tanaka made up his mind to write about his depression in manga.[4] The second influential book for Tanaka was none other than Hosokawa's *Tsure utsu*, which came out as a blockbuster in 2006. The extent to which her manga impacted Tanaka's creation is obvious, as he mentioned the title frequently and even analyzed its commercial strengths in the interview.

Since Hosokawa's bestseller in 2006, quite a number of manga about depression have been published. Among many graphic memoirs of this sort are Hayashi Takeru's (2007) *Utsubyōki* published by *Medical Review-sha*, Kita Yoko's (2008) *Utsu to tsukiau* by Seiwa Shoten, Inagaki Misao's (2010) *Fūfu de utsurun desu* by Shonengaho, and Abe Yuki's (2013) *Watashi wa hataraku utsu wūman* by Shogakukan. However, what separates Tanaka's (2017) *Utsunuke* from those is that this is the only manga written for fellow tōjisha by a man who suffered for close to ten years, which is a fairly long period to be a chronic patient.

The Author's Background

Artist Tanaka is a rare bird in the manga industry in three aspects. First, his drawing style is an adaptation of the famous *mangaka* Tezuka Osamu's artistic characteristics, in particular, the way the characters and *kakimoji* (handwritten words) were used. Second, he is one of the male manga artists who have published autobiographical manga, a minority at this point in time. Third, he has two careers, both as an established *mangaka* and as a corporate employee, which enabled his *Utsunuke* to touch upon the intersection between work stress and depression. At the start of our interview, I was surprised to see the phrase "manga artist specializing in dirty jokes and vulgar parody" (*shimoneta ogeretsu parodi mangaka*) on his business card that he handed me. *Utsunuke* is a serious story about depression with no indecent drawings or obscene language. Furthermore, although Tanaka is a professor of manga studies at Kyoto Seika University, the business card mentioned nothing about that status. This section describes how this self-proclaimed "vulgar" gag *mangaka* came to illustrate his experience of depression in a thoughtful yet entertaining manner, using a variety of innovative visual and textual metaphors for the mental illness.

Tanaka was born in 1962. As a university student, he made a debut at age 22 as a mangaka with his first comic work, *Mr. Coward* (published in the magazine *Komikku gekigajuku*), in 1984. The genre in which his manga career began was *gag manga*. Upon graduating from the university, he was employed at a toy company. His dual career—weekday "salaryman" (salary-earning corporate worker) and weekend *mangaka*—began, and he kept up with the demands of both occupations even during the hectic time of the bubble economy (1986–1991). After working as a salesperson for the toy company for ten years, he switched to the industry of game software makers while continuing to produce new manga. His notable works include *Dr. Chichibuyama* 1986, a *yon-koma* (four-panel comic strip) erotic manga published as a comic book in 1987 as well as *Shinbatsu* (*Divine Punishment*), his first "parody manga"[5] which adopted Tezuka's artistic characteristics published in 2002. Because of this past, his business card has the "dirty-joke and vulgar parody" statement. His first interview-style manga, *Pen to hashi* (*The Pen and Chopsticks*), was published in 2017, just a few months before his bestseller, *Utsunuke*, went into print. Currently, he is employed at BookLive, a digital book sales company; teaches part-time at Kyoto Seika University; and is finishing his most recent manga, *Wakage no itari* (*Youth Is not Wasted on the Young* 2019), a collection of episodes about breakthroughs made by young engineers of a game software. Upon reviewing the resume of such a highly productive person, one would be incredulous to learn that Tanaka developed his depression around the year 2005 and went through relapses for close to ten years.

Tanaka humbly describes himself as an average, run-of-the-mill kind of manga artist. "I didn't graduate from an art college, nor did I formally study drawing. I just learned to draw on my own," he said. From his point of view, graduates of art colleges would have studied the basics, practiced critical exercises such as sketching (*dessan*), and been trained to perceive objects in the three dimensions of length, width, and height. Similarly, he went on, folks who have been drawing manga since childhood would be conditioned to perceive their surroundings like an artist, seeing "outlines" that shape material even though such outlines do not exist. He explained, "I didn't have that sort of perception. So, I had a real hard time working in this profession." According to him, a skillful manga artist can draw anything in black and white with the exception of gray created with screentone sheets. During our interview, this college professor/manga artist continued with his lecture-like explanation, citing *Kodoku no gurume* (*Solitary Gourmet* 2020), a popular manga written by Kusumi Masayuki and illustrated by Taniguchi Jiro. "One amazing characteristic of that [monochrome] illustration is [Taniguchi's] ability to have the reader imagine unseen colors, such as the green of green tea. Of course, people in Japan see green tea every day. So, in their mind, a green tea schema exists, consisting of its typical teacup shape, typical position on a lunch tray, and so on. The tea is drawn in a way that validates the schema of green tea, not that of black tea. Only skillful manga artists can accomplish that!"

Tanaka did have some art training prior to his debut, however; he spent at least a year in a private school, Gekigasonjuku, in Kobe founded by the prolific manga author Koike Kazuo in 1983. On a talk show (Netolaboenta 2016), Tanaka said that he had been making manga drawn in *gekiga*-style and with dirty-joke gags (*shimo neta gyagu*) until, one day, he was told by the editors that his style had fallen out of fashion. That was the beginning of his adaptation of what he calls the "Tezuka-touch drawing" (which I term "Tezukaresque" hereafter). Using his innovative, Tezukaresque technique, he published *Divine Punishment* in 2003. That is why his current characters are round and cute, similar to Tezuka's, while his storytelling style follows the mode of gag manga, the genre in which Tanaka's manga career commenced some thirty years ago. Tanaka told me, "Gag manga's main goal is to make readers laugh. So, you have to stay emotionally close to readers, constantly looking for their tickle spots." Tanaka is reported to have said in a different interview (*President Online* 2019), "When I was a single-skilled gag *mangaka*, I was compelled to tell funny stories and had to try outrageous techniques that nobody had attempted. Gag manga artists are like top-level athletes who are pressured to keep setting new records while their minds and bodies are worn out."[6]

After his long career in gag manga, Tanaka ventured out of the genre for the first time by publishing a serious comic, *The Pen and Chopsticks* (*Pen to hashi*), in 2017 (*President Online* 2019). At first, he was resolved to stick to the gag manga genre because he was a "comedian (*owarai-geinin*) by nature," even though he was a bit envious of the success of some peers who had crossed the border. However, contrary to that resolution, he transitioned to a serious narrative because the content of *The Pen and Chopsticks* did not fit the gag manga style. It was followed by *Utsunuke* and *Wakage no itari* (*President Online* 2019). He eliminated clowning gags and succeeded in straight storytelling with more confidence, writing *Wakage no itari*. After exiting the depression tunnel, he entered his most productive time, both physically and mentally (*President Online* 2019).

When did Tanaka begin identifying himself as a parody manga artist? Tanaka told me that the field of parody manga is a "gray zone," an area in limbo where one can be accused of copyright violation if not done properly. He also taught me that the other label of parody manga is *itako manga*, in which the word *itako* refers to professional mediums who are blind women residing in northeastern Japan and are trained to communicate with the spirits of the dead. In a metaphorical sense, *itako mangaka* are the manga artists skilled at recreating an "X-like" drawing style, as in necromancy. "Initially, we [*itako mangaka*] were messing around privately, pretending that we had been possessed by the spirits of the deceased virtuosos. That play has gained popularity in Twitter and is now turning into a new genre," he explained. He described the following aspects of this emerging field: one must obtain permission from the surviving family (Nanaka has done so with Tezuka's daughter, Tezuka Kurumiko); the practice should entail expanding upon the original drawing (e.g., creating a style as "He would draw like this if he were still alive"); one must avoid acts of copyright impingement, that is, partializing the artist's characters (e.g., drawing the very character of Astro Boy as opposed to drawing a new character by applying the visual codes of Astro Boy).

He entered the field of parody manga with *Shinbatsu* (*Divine Punishment*), a collection of stories with derivative characters spoofed from classic manga titles such as Tezuka's *Mitsume ga tōru* (*The Three-Eyed One* 2010) which began to be serialized in Kodansha's magazine in 1974. The manga originally appeared in *Comic Cue Vol. 6 Tezuka Osamu Remixes* in 1999 and was later printed as a *takōbon* by East Press in 2002, for which Tanaka had obtained permission in advance from Tezuka Production Company (*Netolaboenta* 2016). With *Shinbatsu*, Tanaka first experimented with his iconic Tezukaresque drawing by means of imitating the star *mangaka*'s *kakimoji* (handwritten text) and drawing techniques in developing stock characters. In a magazine interview (*Netolaboenta* 2016), Tanaka revealed that he used the complete collection of Tezuka manga titles owned by his ex-wife to carefully analyze the artist's style and train himself to reproduce its characteristics a full decade after Tezuka unexpectedly passed away in 1989. It was Tanaka's way of preserving Tezuka's legacy, which no one had attempted earlier.

A manga studies scholar, Otsuka Eiji (2017), states that Tezuka's characters were based on round, Disney-style cartoon characters and were developed through the application of various manga semiotic signs. In other words, once an artist deciphers a combination of signs used to create a certain character, it is technically attainable to reconstruct it the way Tezuka drew. According to Tanaka, "Master Tezuka had certain rules for drawing. Once you master the rules, you can recreate his manners. For example, his characters tend to have a round face, and their mouth and chin or mouth and nose are drawn relatively close together. He also used a lot of shades of grey in his colored manga" (*Netolaboenta* 2016). In preparing *The Pen and Chopsticks* several years later, he learned to reproduce several other famous artists' styles. In a magazine interview (*President Online* 2019), Tanaka stated that working on this manga was a great way to learn the master-level drawing techniques since he had to skillfully imitate a wide range of drastically different styles invented by star *mangaka* including Tezuka Osamu, Akatsuka Fujio, Chiba Tetsuya, and George Akiyama. However, when writing *Utsunuke* (as well as *Wakage no itari*), Tanaka utilized only the Tezukaresque drawing technique. He told me that he had chosen this style because Tezuka's art is "evergreen [*ebāgurīn na e*]" and "reader-friendly." To Tanaka, both Disney and Tezuka cartoons are ageless models that stand the test of time.

As mentioned earlier, Tanaka is a depression *tōjisha*. While working as a salesperson at a game software company, he began to experience several depression symptoms around 2005. It was not until 2006, however, that Tanaka finally sought medical help. At age forty-five or so, he visited a mental health clinic for the first time in his life and began taking, Depas, a medication used to treat conditions such as depression, anxiety disorders, and sleep disorders (*Gendai Business* 2018). However, the medicine stopped working for him, and he switched to another doctor, only to end up with stronger antidepressant pills. He changed doctors a third time, went back on Depas, and commenced his recovery from depression thereafter (*Gendai Business* 2018). He told me, "At first, I thought it was 'male menopause' or something. I didn't have the faintest idea, as I didn't believe I was the type of person who would get depression." What type of person did he think would develop that illness? In an interview article (*Gendai Business* 2018), he argued that personality types that are susceptible to depression are people who are unplayful (*kimajime*) and perfectionists (*kanpekishugi*) who cannot settle for a moderate degree of commitment (*tekitō de iikagen ni dekinai taipu no hito*). Once he allowed himself to listen to an inner voice saying, "I hate doing this," he began to feel relieved, he asserted; thus, he encouraged others to go easy on themselves (*amayakashi te ii*). However, it should be noted that no personality is scientifically proven to cause this mental disability. We must be careful not to single out any "type of person" and inadvertently create a stereotype.

According to an episode in *Utsunuke* (p. 2), Tanaka was working practically twenty-four hours a day, seven days a week as a full-time salaryman and a manga artist. However, Tanaka's character insists that overwork was not a direct cause of his depression, saying, "Depression set in because I began to hate myself (*jibun wo kirai ni nattakoto*)" (p. 5). Tanaka provided an expanded account of his health history in a magazine interview (*Gendai Business* 2018): soon after he was employed at the toy company, Takara, in 1984, he was transferred to the headquarters in Tokyo, which was "a super 'Black' workplace."[7] In 1996, he developed hyperthyroidism due to work stress, according to Tanaka, although it should be noted that the direct link between stress and this illness has not yet been scientifically proven. His thyroid issue was quickly managed with the proper medication, and around that time, he began to experiment with the Tezukaresque drawing style for his latest manga project. In that interview, he further pointed out that his depression might have been avoided, had he changed his stressful job sooner (*Gendai Business* 2018).

Tanaka's depression symptoms included having a depressed mood all day long, feeling anxious, and having an inexplicable feeling of being miserable (*tsurasa*) day after day (*Gendai Business* 2018). Similarly, in the first episode of *Utsunuke* (p. 2), Tanaka's character is depicted with a dull and glassy-eyed look. The accompanying text explains that he feels as if his body, soul, and brain had all turned gray, having diminished pleasure with any of the music, films, or landscapes he used to enjoy. The character's fixed expression indicates that he experiences debilitating anxiety, which eventually leads to a strong suicidal ideation, particularly when he is severely depressed. The character tells the reader that he planned to end his life before his fiftieth birthday (pp. 1–3).

Fortunately, Tanaka's condition began to improve after he adopted the morning ritual and method of mind training recommended in Miyagima's book (described in "The Title and Target Readership" section). "The book taught me that I must come to like myself to have a chance to exit the [depression] tunnel," he stated (*GQ Japan* 2019). Fortunately, by 2012, he was able to discontinue the medicine and gain control over his life (*iikanji ni natta*) (*Gendai Business* 2018). As illustrated in *Utsunuke*, Tanaka's character notices that his sense of pleasure is gradually restored about two months after he began practicing the self-affirmation method; he then becomes able to engage in new projects and activities (pp. 14–15).

However, after he gets laid off from the game software company where he did not enjoy working and is happily working at a new company, ironically, his depression comes back (p. 17). Tanaka's character calls the relapse an "abrupt return" (*totsuzen ritān*), which is visually represented with a bunch of dark, little, wiggly creatures (p. 19).

Among his depression symptoms, what Tanaka dreaded most was a panic attack accompanied by an inexplicable sense of anxiety—fear and worry of an unknown origin—which was nothing like what he had experienced before (*GQ Japan* 2019). Cleverly, he came up with an idea to observe and record his daily condition in a journal and eventually figured out the role of climate: a sudden change in temperature (*hageshii kionsa*) seemed to trigger his depressed mood. The data he collected enabled him to get a bird's-eye view of his health (*jubun wo fukansuru*) and to predict his relapse patterns (*GQ Japan* 2019). In *Utsunuke*, by making a graph of daily temperatures along with a note of depressive symptoms, Tanaka's character discovers that sharp drops in temperature that typically occur in March, May, and October somewhat correspond with his mood swings (p. 20). It should be noted that seasonal affective disorder (SAD), a condition of developing depressive symptoms and anxiety during the winter, does not account for Tanaka's depression relapses since his "tough" months are not in the winter per se.

The discovery enabled Tanaka to anticipate the next relapse, making the abrupt return less abrupt. To him, that moment was like a fog finally lifting in front of him, and in *Utsunuke* (p. 23), his character comes to realize that depression is not like a shape-shifting *ūrei* (spirit) but is more like a *yōkai* (monster) with predictable characteristics. As he picks up a couple of creatures, he utters, "That's right. These are not *yūrei* but *yōkai*. Even though I don't welcome them, I don't have to fear them as long as I know how to deal with them" (p. 23). This realization also helped Tanaka accept *kankai* (remission), not *kanchi* (cure), with the possibility of relapses. By October 2014, he had successfully "exited" his depression tunnel, and the serialization of *Utsunuke* started in December of the same year through an online manga reading site, *Digital Shosetsushi, Bungei Kadokawa* (*GQ Japan* 2019). He said to me, "We can't make light of our brain's health."

Tanaka's Drawing Styles and Storytelling Characteristics

Tanaka's artistic skills to visualize depression symptoms are exceptional. His strategies to represent psychological states that are invisible to outsiders include the personification of depression, embodiment of happiness, and representation of the faceless public. That strength alone makes *Utsunuke* an exemplary *tōjisha manga*. This section elaborates on a variety of innovative visual and textual metaphors that he employs in this manga to illustrate his own (and others') lived experience of depression.

Visual Metaphor of "The Depression Tunnel"

In *Utsunuke*, the metaphor of "*utsu tonneru*" (the depression tunnel) is visually represented in one of the two methods throughout the twenty short episodes. As stated earlier, this tunnel image is a recurring theme of this manga, expressing the tōjisha's feeling of entrapment. One representation is a hole-like exit of a black cave, as seen in the bottom right panel of page 8 in Fig. 7.2. Tanaka said, "If you're inside, it feels like an unpeaceful void. So, from an inner perspective, it rather feels like a cave made of sharp icicles and ragged walls."

He also draws a tunnel with a train-track tunnel (see bottom panel of Fig. 7.3), which is "an outer perspective" into the state of depression, according to him. The metaphor of a dark tunnel-like conduit through which one travels like a train, in particular, resonates with non-Japanese who are familiar with the English-language expression, "There is light at the end of the tunnel." However, considering Tanaka's main purpose of writing *Utsunuke*—to remind those who feel stuck in this tunnel that it has an exit—it is clear that the emphasis is on the "end" of the journey, whether the confined space is a cave or tunnel.

Personification of Depression: Utsu-kun

The most iconic representation of depression in *Utsunuke* is a bundle of little creatures called *Utsu-kun* (see middle panel in Fig. 7.4), which sounds like a little boy's name. Nagano (2019), a patient at the Bethel House in Hokkaido (discussed in chapter "Tōjisha"), explains how naming one's own mental health problem can be an effective coping mechanism. As depicted in his illustrated essay, Nagano named his depression *Fuan-san*

Fig. 7.2 Image by Tanaka Keiichi. (Courtesy of KADOKAWA)

Fig. 7.3 Image by Tanaka Keiichi. (Courtesy of KADOKAWA)

(Mr./Ms. Anxiety) and holds an imaginary conversation with the personified illness on a regular basis. Nicknaming one's own condition, he argues, helps create some distance from the psychosis and enables the person to deal with it with much less emotional involvement (p. 11). He states that that sort of objectivity prevents the sufferer from being submerged in the illness and makes it easier to talk about the experience publicly.

In Tanaka's interpretation, *Utsu-kun* are "super friendly" beings, despite their ironical role of bringing him psychological torment, so these creatures are drawn like small, whitish animals in most scenes (e.g., Fig. 7.4). They also appear as black, slimy, sludge-like creatures in some scenes (e.g., Fig. 7.5). Tanaka told me, "The unique visualization of depression, I heard, made this manga appeal to many readers. It helped the readers, especially those who had never experienced depression themselves, understand how it affects the lives of others. That's why the manga sold so well." He believes that such a representation of mental illness is only possible in a visually oriented art form like manga because words alone "probably could not express it transparently."

How did Tanaka come up with that particular strategy of personification? He responded,

> While brainstorming with Ms. Kaneko [his editor] at the beginning, we reviewed Azuma Hideo's *Shissō Nikki*. Mr. Azuma's roundish characters ('*marukkoi e*,' an equivalent of *chibi*-characters), in contrast to the awful story of his suffering mentally and living as a homeless man, resembles my own [Tezukaresque] drawing. He tells the tale in a masterfully nonchalant way with the presence of those iconic round figures, which helps make the heartbreaking story digestible. I told Ms. Kaneko, 'We need to aim at that ambience.'

Tanaka understands that Azuma, a fellow gag mangaka who passed away recently, excelled at softening the edge of a potentially brutal theme with his cute *chibi*-style drawing. "In the process, my slime-like characters, *Utsu-kun*, came to look like Mr. Azuma's peculiar 'depression monster.' His technique definitely influenced mine."

Unlike Azuma's monster, however, Tanaka's depression is represented seemingly in multiple forms: white minions (e.g., Fig. 7.4), black slimes (e.g., Fig. 7.5), and at one time, a Godzilla-like monster (p. 156). I asked Tanaka about this varying embodiment of the same illness. He responded, "Actually, the shape is the same; I only varied the colors. Another trick I used is drawing them in different manners, such as being piled up *en masse* or dangling from above." In some scenes, *Utsu-kun* are shown flying away

Fig. 7.4 Image by Tanaka Keiichi. (Courtesy of KADOKAWA)

Fig. 7.5 Image by Tanaka Keiichi. (Courtesy of KADOKAWA)

(first panel in Fig. 7.6). Regarding this image of "flying" white minions, Tanaka said that their whiteness represents the state of depression being "purified" before fading away. Tanaka summarized succinctly in a magazine interview (*GQ Japan* 2019): *Utsu-kun* are round, white creatures that gather in mass or sometimes turn black but fly into the sky when the person gets out of depression.

"Either way, Mr. Azuma's drawing has a significant impact [on Tanaka's drawing]," he told me. Another influential artist might be Miyazaki Hayao. Right below the flying creatures in the first panel of Fig. 7.6 is handwritten text, stating "Are they *Makkuro kurosuke*? No way!" *Makkuro kurosuke* is a small, black sprite-like figure in the shape of a hairy ball that appears in Miyazaki's anime, *My Neighbor Totoro*. It is a common practice for mangaka to cross-reference well-known illustrations within their works, and the scene of *Utsu-kun* creatures flying away into the night sky in that panel might simply be a homage to that anime.

Embodiment of Happiness: Foot Warmer and Greenfield

The polar opposite of the dark, cold depression tunnel are the symbols of happiness that Tanaka embedded in this manga. A sense of well-being is signified by the landscape of a greenfield, as well as the image of a *kotatsu* (Japanese electric foot warmer), which comically props up in the field. These images usually show up at the beginning of an episode (e.g., bottom left panel on p. 32 in Fig. 7.6). "The *kotatsu* is a visual metaphor for the refreshing feelings after getting out of depression," he described. But why the foot warmer? He replied that, to him, the *kotatsu* is the most iconic appliance in the Japanese household, and Japanese people are familiar with the feeling of happiness when they come home in the shivering-cold winter and snuggle into a warm *kotatsu* in their living room. Thus, he drew the image simply to evoke the sentiment of euphoria. "By contrast, my trick of depicting the depressive mood is drawing a dark underworld with rugged, sharp rocks," he added. The *kotatsu* is often placed in the background of *sōgen* (greenfield), which is meant to symbolize spring. Together, the opening scene of each episode is always a place of warmth and happiness. "That's my intention," he said.

Representation of the Faceless Public: Assistant Kaneko

Kaneko-kun is a male character who wears an eyeball hairband and poses as Tanaka's assistant in *Utsunuke* (Fig. 7.7). At the beginning of each episode,

Fig. 7.6 Image by Tanaka Keiichi. (Courtesy of KADOKAWA)

Fig. 7.7 Image by Tanaka Keiichi. (Courtesy of KADOKAWA)

Kaneko-kun introduces himself, saying "I'm his assistant, Kaneko, but have nothing to do with depression" (e.g., *Utsunuke* p. 4) or "I'm Kaneko and don't understand what it's like to be depressed" (e.g., *Utsunuke* p. 18). The name of this assistant may be derived from Tanaka's real-life editor, Kaneko Akiko, although she appears separately as the featured character in Episode Five with the pseudo-name Ori Haruko. In real life, Ms. Kaneko has experienced depression, too, and Tanaka told me, "The manga would have not been this successful without her collaboration."

Although it is unclear what his eyeball hairband symbolizes, it is self-evident that the character represents ordinary people whose lives have not been affected by a serious mental illness problem—the faceless public. The character's offbeat comments about depression are reminiscent of what lay Japanese people might say to the afflicted. His questions also lead Tanaka's character to impart critical information about his condition or the characteristics of depression in general in each episode. Furthermore, *Kaneko-kun* sometimes expresses the voice of opposition, saying things such as "That's such a cliché" (*Utsunuke* p. 13) and "His story is all lies, isn't it?" (*Utsunuke* p. 18). Thus, he appears to play the role of *tsukkomi* in the standard comedic format (discussed in chapter "Essay Manga"). I believe that the presence of this character is a tactful consideration in terms of storytelling because many readers with no direct experience of depression probably find his thinking and commenting to be very relatable.

Brain Fog: Gelatin or Muddy Agar

Another clever visual gimmick Tanaka invented is what he calls *nigotta kanten* (muddy agar), similar to gelatin. To depict the state of brain fog, Tanaka drew a translucent substance that covers the brain as shown in the middle panel in Fig. 7.7. The image resembles a traditional Japanese sweet, *mizuyōkan* (sweet azuki beans covered with agar, or seaweed gelatin). To the right of the drawing of the muddy brain, he writes that this symptom is "like the brain being constantly addled with a haze" (*Utsunuke* p. 11). Tanaka said that, when he had brain fog, his sensory system responsible for processing new information did not function, so he felt as if his brain were covered with some muddy agar. "That [*nigotta kanten*] metaphor is based on my direct experience of brain fog," he told me.

The experience of brain fog is a common symptom of depression, affecting 80–90% of the depressed. It is also termed *cognitive dysfunction* (CD). It affects the person's mood and ability in a variety of ways related to mental processing. One of them is the inability to concentrate, which leads to the inability to finish a task at hand, follow a conversation, or complete a thought. Because of brain fog, one may have trouble focusing on what they are reading or a movie they are watching. Another consequence of brain fog is having trouble with one's memory (e.g., being unable to recall what they were doing moments ago). The person may also experience difficulty with executive functioning (e.g., unable to make decisions or perform multitasking). The person's reaction time is affected, too, as if the brain is blocked. The condition may continue even after the depression is in remission; it can lead to a more severe form of depression. However, CD can also be caused by other factors such as pregnancy, menopause, cancer chemotherapy, lack of sleep, and other conditions and illnesses, including COVID-19. Therefore, it is important to state that CD is not a depression-specific symptom.

Use of Powerful Phrases

During our interview, Tanaka explained the value of using powerful phrases, which he calls "power words" (textual metaphors), such as *utsu tonneru* (depression tunnel), *nukeru* (pass through, in reference to "going into remission"), *nigotta kanten* (muddy agar, in reference to "brain fog"), and *totsuzen ritān* (depression relapse). The phrase that is most distinctive of Tanaka's textual metaphors may be that depression is a "cancer of the mind" (*kokoro no gan*) exemplified by the bottom left panel in Fig. 7.8. The most common metaphor for depression in public discourse is actually "the mind cold (*kokoro no kaze*)." Contrary to this public symbolism, Tanaka presents this condition as a lethal ailment of the *kokoro* (mind, spirit, heart), like a cancer, which is far more pernicious than a mere cold. He even writes in the manga, "Suicide is one of the symptoms of this cancer" (*Utsunuke* p. 151), implying a close connection between depression and death similar to cancer.

Historically, the "cold" metaphor was used in pharmaceutical commercials to promote the sales of selective serotonin reuptake inhibitors (SSRIs)[8]: just like a cold, this medicine can treat it. It was also intended to reduce stereotypical views of the illness such as "only weak-minded people get it" by emphasizing the fact that depression can happen to anyone (*dare ni demo kakaru*). Naturally, the "cold" metaphor was used to refer to depression in graphic memoirs like Hosokawa's *Tsure utsu* and Inagaki's *A Depression Couple*. Tanaka applies an analogy of cancer, instead. It was a

Fig. 7.8 Image by Tanaka Keiichi. (Courtesy of KADOKAWA)

phrase his editor Kaneko found in the interview transcripts and proposed to use as a powerful textual metaphor in his manga. Indeed, depression can be far more serious than having a cold, especially because of the close link between this illness and suicide. In 2006, a suicide prevention law was passed in Japan. Beginning in 2015, workplace stress checks began to be integrated into company practices. A typical stress check is a questionnaire composed of items related to symptoms of stress. Taking this check is now mandatory for companies with more than fifty staff members; smaller companies are only "encouraged" to do so, however.

Another example of textual representation that is equally impactful is "fleeing from work" (*shigoto kara nigeru*) in the Terumi episode ("What worked is his [Terumi's] 'fleeing from work' after all" *Utsunuke* p. 32). In a magazine interview (*GQ Japan* 2019), Tanaka stated that, just as in the Uchida Tatsuru episode, only after realizing that "it is perfectly okay to rest" can we finally set ourselves free and begin to get better. He urges readers to permit themselves to evade responsibilities when their situation taxes their mental health. Relating to his personal history, he considers it wise to leave the source of stress, whether it is a toiling relationship with your superior or colleague or a mismatch between your profession and your disposition.

The Effect of Colors

Unlike regular, black-and-white manga, *Utsunuke* is in a multicolor format. "It [the *takōbon* version] was digitized from the beginning because it was never published on paper," Tanaka explained. The original story came out on both Bungei-zasshi, KADOKAWA's digital site, and Note (https://note.com/keiichisennsei), a social media site where members are allowed to post their comic works. "Both places are websites. Web manga is, nowadays, usually in color."

Another reason for *Utsunuke* being in color is Tanaka's own preference. He told me, "After all, the use of colors helps the author make up for his artistic shortcomings. Even a poorly skilled person can draw a 'cloudy sky' if grey is added or a 'sunset sky' if painted in orange, you know." He explained that drawing a monochromatic manga poses an extra hurdle because one must distinguish gray, blue, and orange skies with black-and-white brushwork only. By contrast, a polychromatic manga allows the author to readily make that distinction through the added element of color.

His third reason appears to be indicative of his powerful creativity: to reproduce the two contrasting mindsets—being deeply depressed and feeling well again—by taking full advantage of colors, which are available in digital format. "I use color modulation to my advantage. For example, I paint a background in almost black and white to depict the depressed mood. Or I use bright colors to express a blue sky after coming out of depression, and so on."

Storytelling Strategies

Besides his tactful strategies inventing these visual and textual metaphors, what makes Tanaka's manga special is its intriguing story development. For example, to present Tanaka's own story of depression, he begins the first episode with a hypothetical funeral scene in which Tanaka's character is already in the coffin, surrounded by roses. On the very first page, his character narrates: "May 4, 2012, could have marked the day I, Tanaka Keiichi, died" (p. 1). With that alarming beginning, the reader is likely to feel compelled to read the first three episodes to find out what happened to Tanaka. The rest of the seventeen episodes are told in a similarly entertaining manner. For example, a former office worker shows up in sheep's clothing, evoking a Japanese term, *shachiku* (literally, "company livestock" to mean "company slave"), which visually associates her life with that of a livestock animal. The character then tells her experience of depression by overwork forced by her employer. Tanaka explained, "Such tricks are important because they help lighten up the heavy weight of the theme, depression." As a storyteller, he wanted to entertain the reader of *Utsunuke* with inspiring episodes and amusing images.

Objectivity and Humor

During the interview, Tanaka often emphasized, "As a *gag mangaka*, I have to make the reader laugh." What does he really mean by that? In Tanaka's view, humor helps him attain objectivity. Tanaka went on to explain that creating a good story is only possible after the author succeeds in gaining some distance from his own tragedy and becomes able to examine it objectively. Citing Ueno's manga, *Without Saying Goodbye*, Tanaka told me that Ueno was unable to revise his first draft while still mourning his wife's passing, so he had to wait for a while.

In an interview inserted at the end of *Shissō Nikki*, Azuma Hideo similarly stated that at first, he thought of casting himself as a cat to create distance between himself as the author and his experience of the illness. Azuma told his interviewer that when he was in the midst of depression, it was impossible to have such objectivity. He was overwhelmed with guilt, since his family also went through a painful period. Only after he returned home and thought of making manga about his homeless experience, was he able to look at the underlining mental illness from a third-party perspective. He was able to "laugh" at himself as the character and revisit his miserable experience as the character's story. As a gag mangaka, Azuma used humor to objectify the experience just as a comedian is able to laugh at himself (p. 196).

Tanaka told me that he made sure to embed a funny moment in each episode as a depression survivor in *Utsunuke*. "In every case, the former half of the story is about suffering from depression and the latter half is a story of 'escape' [from the depression tunnel]. Unless the two halves are interspersed with some funny elements, the reader might get tired of reading it." *Utsunuke* is also infused with the element of inspiration, since Tanaka makes sure that the manga has as much humor as possible, especially if the reader is someone with depression. "People won't read a story that just tells one sad story after another. So, I made a conscious effort [to add humor]. Plus, I'm a gag mangaka by nature and can't help doing so," he smiled. In Japan, there is a stereotype of an Osaka native as a person of entertainment. I asked him, "Does this have anything to do with the fact that you are from Osaka?" "To some extent, yes. [In Osaka] we say, '[*geinin wa*] *warakashite nambo*' (the true worth of an entertainer is his/her ability to make people laugh)," he replied jokingly. Tanaka's conscious effort of humorous storytelling is obviously a significant factor of *Utsunuke*'s popularity, and laughter seems to be the adrenalin of his creativity.

Reception and Generalizability

Utsunuke's strong, positive reception is readily apparent, both in terms of its sales and adaptation into a television drama series viewable on Hulu Japan in 2018. In addition, *Utsunuke* was a top ten finalist, nominated for the 2018 Tezuka Osamu Cultural Award in the manga division. Regarding the fact that more than 340,000 copies have been sold, Tanaka said, "I had no idea!" as if he were still incredulous. His previous manga titles created in gag manga and parody manga had never sold as much since those genres have only a small target audience. To that point, "I had been a [manga] author whose track record was 10,000 copies, more or less," he said. "I was almost certain that my style would not generate more than that."

However, Tanaka had a hunch that *Utsunuke*'s sales might be a bit different because, first, he aimed at a wider range of readers—from the depressed to those who knew someone with the illness or who just wanted to learn about it. Second, knowing that Hosokawa's *Tsure utsu* had been a megahit several years earlier, Tanaka was hoping for 30,000–50,000 copies although he desired twice that number. "I was telling others, '100,000 would be great.'" To be called a "bestseller," a book must sell more than 50,000–100,000 copies according to a publication sales promoter, Nishiura Koji (2019). However, Nishimura warns that in today's recession in the publishing industry, the majority of new books (including manga) rarely hit those numbers, regardless of genre. Takemiya Keiko, a seasoned *mangaka* and former chancellor of Kyoto Seika University, also states in her book (2014) published with Uchida Tatsuru that the manga industry is "a harsh group where one becomes a legitimate member only after having sold, at least, 100,000 copies" (p. 73). Tanaka's lifetime goal was to publish a manga that would sell 300,000 copies or so. Luckily, with *Utsunuke*, his goal has been met as more than 340,000 copies of the book have been sold. He commented, "I'm proud of the number as a clear indicator of commercial success."

The prior commercial success of Hosokawa's *Tsure utsu* (and the readership base created by it) was partially responsible for the strong audience reception of *Utsunuke*, according to Tanaka. As mentioned in the section on target readership, besides Tanaka's target of patients with depression, both families affected by the illness and ordinary people in good health who are curious about it also bought *Utsunuke*. Immediately after its publication, Tanaka learned from reader feedback that *Utsunuke*'s visualization of this mental illness (e.g., *Utsu-kun*) was so unique that it set the book apart from other, similar manga titles. As he stressed earlier, "Visualizing the invisible is a great advantage of manga over a text-only medium." Another characteristic of *Utsunuke* that separated it from the rest is its addition of sixteen other cases to his own memoir, Tanaka asserts. More detailed analysis of how this aspect enhanced the manga's generalizability will be provided in a later section of this chapter.

Furthermore, Tanaka shared the commercial strategies that were chiefly responsible for the manga's success. "I've already disclosed this in many interviews: it's not my single power that made *Utsunuke* an instant success," Tanaka said at first. Then, he described how the success was the result of his collaboration with his manga editor, the *takōbon* designer, and the publisher's public relations department. "I cannot overstate the fact that exceeding 300,000 copies in sales was possible only with these three talents working for me," he commented.

As mentioned earlier, Kaneko Akiko was his *tantō henshūsha* (editor assigned to the project). In the epilogue of *Utsunuke* (p. 173), Tanaka wrote, "Had I not teamed up with you, Ms. Kaneko, I don't think I could have made *Utsunuke*." Before signing a contract with any publisher, Tanaka tweeted a message, "I came out of depression and want to write a manga about it. Would anyone be interested in becoming my *tantō* [*henshūsha*]?" KADOKAWA's manga editor, Kaneko, tweeted back in less than thirty minutes, "Please let me do that." She was extremely interested in Tanaka's project because she too suffered from depression. Tanaka also felt that, due to that shared experience, she would be a great partner. "She understands the agony of the illness. That's why we got along so well," he told me. In addition to being an editor for various manga artists, Kaneko also worked in the field of arts and literature (*bungei*), which is the genre of her true passion. "All that a novel has is text, you know. No graphics. So, writers put their lives on the line with their choice and use of words. [Working in the field of *bungei*], she's cultivated an exceptional ability to choose the right word in the right context," Tanaka continued. Kaneko helped extract such impactful words as "the cancer of the soul," "gain back your healthy narcissism [*kenkotekina narushizumu wo torimodosu*]," and "consider it [depression] as a time for self-examination [*jinsei no jishu jikan dato omoe*]" from the seventeen interviews Tanaka and Kaneko conducted together for the episodes. "She sat [with me] through each interview, which would go on endlessly, like two hours or so. After that, she would review and pull out a thematic phrase that highlighted the whole interview." She then emailed Tanaka an interview summary, an A4-sized, one-page paper, together with the thematic phrase(s). Tanaka drafted a plot from that theme and discussed the draft at their next meeting. He added, "Because of her editorial experience in the field of *bungei*, such a process was possible" and "Her advice was pertinent. So, I trusted her sense, thinking 'She's really good at it.'"

Another influential person was the designer of the takōbon who worked in the non-fiction division of KADOKAWA. That person had selected an eye-catching illustration from all of the images of *Utsunuke* and placed it on a pinkish background for the book cover, making it a customer-friendly product, especially for female customers. For the book's strap, which tells customers what it is about, the designer featured his drawings of the interviewees, most of whom were recognizable people in popular culture and other fields, along with a message that summarized the episodes, "This is how we came out of depression." (The strap identified *Utsunuke* as an essay comic, however.) "That strap cleverly informs the reader that this is a manga, an essay comic, about how to escape depression based on interviews with famous people. It's all clear with this strap." The designer worked in the field of non-fiction writing. "Non-fiction is the genre where books don't sell just for who the author is. It takes a good title and content as well as its cover that succinctly tells the customer what the book is about. My [*takōbon*] designer was someone who understood that intuitively," Tanaka explained. "That's why readers are attracted to a non-fiction like "10 Habits of Successful Salarymen" (*Seikōshita sararīman ga yatteiru jū no shūkan*). A crystal-clear title like that is very important, isn't it?"

The publisher's public relations department was the third key entity to which *Utsunuke*'s success is attributed, according to Tanaka. "Indisputably, the efforts of the public relations department were super effective. They brought to me a wide variety of book-promotion opportunities, from web interviews to those by magazines, newspapers, radio programs, and television networks." He himself responded energetically by attending as many book promo opportunities as possible. The department also tactfully selected the time that *Utsunuke* went into print. "They were clever to pick the end of January for its release. When it comes to depression, it typically worsens in February and March," Tanaka added. As he explains, KADOKAWA, a publisher that also produces movies, knows strategic market release times. "All in all, the sale of some 300,000 copies was a sign of victory resulting from the public relations strategies, rather than my manga's content," Tanaka said humbly.

In addition to these factors of success, the relatability of the manga's topic, depression, is a reason for its strong reception. However, Tanaka understands that a single case is not generalizable. In the first three episodes that depicted his journey with this illness, his character, Tanaka-san, tells the reader, "This is only my case, mind you [*Akumade 'boku no bawai wa' desukedo*]" (p. 14). Tanaka himself stated in an interview (*GQ Japan* 2019) that this manga does not claim to offer a single cure to readers; rather, it was written in the hope that readers will find applicable one of the seventeen approaches to recovery. He explained that readers are likely to lose interest if the book presents only one approach to combating depression and ignores other viable options, like many books on depression on the market. "I thought it'd be useful to investigate various cases of depression and present findings as a layperson. That was the main purpose of my book, which worked well." That is why *Utsunuke* covers a wide range of depression-related conditions including a person who experienced many relapses and another person who was eventually diagnosed with bipolar disorder (one aspect of which can be depression).

The presentation of multiple voices helps increase relatability with the manga's characters. Because no homogenized depression model exists, it is a clever choice. Tanaka's *Utsunuke* presents different cases of interaction between depression and the corporate culture, including Zunzun, who developed depression from overwork, and Tanaka himself, who claims that his work, especially making manga, helped prevent his condition from worsening.

Another reason for having other individuals' stories in this manga is a concern that he and the editor Kaneko had: the story may not last beyond the third or fourth episode if it centers around Tanaka's experience alone. They figured that it would be meaningless to dilute his story to generate multiple episodes. "To make a [manga] book, you need about 170 to 200 pages. There is no way I could write about myself for that long," Tanaka told me. He would usually write a short episode of six to eight pages. "That means I'd need about twenty people [to interview], at least." Fortunately, Tanaka remembered his acquaintance, a game software developer, who had recently recovered from depression. He had suggested to Kaneko, "How about finding other people like him and interviewing them? It will be a collection of useful information." That was how they came up with the interview manga format.

I asked Tanaka how he reached out to other tōjisha besides the game software developer. Tanaka replied, "Basically, they were found through my own and my editor's connections." They had contacted various people through their network, including novelists with whom she had worked. "It may sound a bit distasteful, but we thought it necessary to include some celebrities to market the manga to a wider audience." To that end, they managed to get hold of people such as the musician, Ōtsuki Kenji, and the adult video director, Yoyogi Tadashi, both of whom had publicly disclosed their depression diagnoses. "The publisher's name was also helpful," Tanaka added, "because KADOKAWA is a well-established household name in the publication industry." He emphasized that listing those celebrities' names on the book's strap was an effective strategy, as mentioned earlier. Some of the people they contacted, especially other corporate employers, or "salarymen," declined or later withdrew from the interviews due to their concern that their companies might find out about them even if their identities were altered. Tanaka is grateful for all who agreed to be interviewed and willingly shared how they suffered but got out of depression eventually, which he sees as an act of courage that is no small feat. "Talking about their tough days can trigger their depression," he explained. Tanaka strongly hopes that those who are suffering from depression will benefit from his own as well as the other tōjisha's narratives in spite of the stigma attached to this illness.

Notes

1. The actual text that appears in *Utsunuke* is "*Utsu no tonneru no naka de imamo kurushindeiru anata e. Detekitahito ga iru toiukoto—sore wa deguchi ga aru toiu tashikana shōko desu*" (To those still struggling inside the depression tunnel, this report of people who came out the other side proves that the end exists in the tunnel).
2. Tanaka explained that the title *Gurazeni* evokes a famous quote of the manager of the Nankai Hawks: "*Guraundo niwa zeni ga ochitteiru … Zeni ga hoshikerya renshūsee* (The money is on the ground. If you want money, come to the ground and practice)." The authors, Moritaka Yuji and Adachi Keiji, used an abbreviation for the title to hook baseball fans. Partially thanks to the "hook title," more than three million copies of the manga have been sold, and it was adapted into an anime in 2018. Tanaka added that the four-letter word works well with a long title if fans can easily abbreviate it like Hosokawa's *Tsure utsu* (from *Tsure ga utsu ni narimashite*).
3. According to Sugiura (2017), approximately 300 million people are diagnosed with depression worldwide compared to about one million patients in Japan (1–2% of Japan's population of 125 million people). For example, within the year 2014, approximately 1,120,000 patients are said to have utilized psychiatric services for depression or bipolar disorder. However, because Japan has 25,000–30,000 suicide deaths every year, she suspects that there are many more depression sufferers who are undiagnosed.

4. Azuma Hideo, another gag mangaka, also expressed a similar desire ("I shall use this experience as my manga's material" [*kore wo neta ni shiyou*]) in the epilogue of his 2005 manga, *Shissō Nikki* (Disappearance Diary). Although Tanaka did not mention it, he later told me in the interview that he was influenced by Azuma's visual representation of depression. Azuma Hideo's *Shissō Nikki* also became a bestseller with approximately 300,000 copies sold.
5. The term *parody manga* contains the word "parody," which means "to make fun of or recreate what was previously created, as in "spoof."" When used in the contemporary comic industry of Japan, the term refers to a type of comic or anime that imitates the artistic characteristics of a classic manga artist (e.g., Tezuka Osamu, Matsumoto Reiji, Takahashi Rumiko) or a parodied version of the artist's work. It is a form of reverse engineering. Examples of the latter type are *Attack on Titan: Junior High* (*Singeki! Kyojin chūgakkō*) and *Spoof on Titan* (*Sungeki no kyojin*), created as parodied versions of the famous anime, *Attack on Titan* (*Shingeki no kyojin*).
6. The genre of gag manga seems to be a pressure cooker of the manga industry, according to Azuma Hideo, a household name in the field. In a postscript interview, "*Ura-Shissō Nikki*" (2005) (Backstory of *Disappearance Diary*, pp. 200–206), at the end of Azuma's award-winning manga, *Shissō Nikki 1*, he confided to his interviewer, "To remain popular, gag *mangaka* are constantly under pressure to make new gags. That pressure can take a toll on our mental health, little by little. Especially if you take that demand too seriously, you'll be driven to extreme" (p. 204). Azuma also describes the desire most commonly shared among gag manga artists: to please all readers with gags. However, we need to be careful not to jump to the conclusion that the pressure of the genre "caused" Takana's depression, as this illness has a combination of interacting factors that lead a person to the condition. Tanaka himself also stated in an interview (*Hobonichi Techō* 2018), "Manga is the profession where my *raison d'etre* [*jibun no sonzaikachi*] is authenticated. I can say that, without it, my depression could have been aggravated."
7. The Japanese term "black *kigyō* (corporate)" refers to a company that treats its employees like slaves under an exploitative working condition.
8. SSRIs are antidepressants, medicines designed to ease symptoms of depression by increasing the level of serotonin in the brain.

References

Abe, Yuki. 2013. *Watashi wa hataraku utsu wūman* (*I'm a Working Woman with Depression*). Tokyo: Shogakukan.

Azuma, Hideo. 2005. *Shissō Nikki*. Tokyo: East Press.

DSM-5 (*Diagnostic and Statistical Manual of Mental Disorders, 5th edition*). 2013. Washington, DC: The American Psychiatric Association.
Digital Shosetsushi, Bungei Kadokawa. 2014. https://promo.kadokawa.co.jp/utsunuke/?_gl=1*f9fkxw*_ga*MTgyODE1MjMyOS4xNjI0MzMxNjM3*_ga_ZB5RJD6B6P*MTYyNDMzMTYzNi4xLjEuMTYyNDMzMTY2My4zMw.
GQ Japan. 2019. https://www.gqjapan.jp/life/grooming-health/20190522/utsu-nuke-interview-1 (retrieved on June 20, 2021).
Gendai Business. 2018. https://gendai.ismedia.jp/articles/-/56893?page=3 (retrieved on Jan. 20, 2020).
Hayashi, Takeru. 2007. *Utsubyōki* (*A Memoir of Depression*). Tokyo: *Medical Review-sha*.
Hobonichi Techō. 2018. https://www.1101.com/store/techo/ja/magazine/2018/utsunuke/index.html?device=pc (retrieved on November 14, 2019).
Inagaki, Misao. 2010. *Fūfu de utsurun desu* (*A Depressed Couple*). Tokyo: Shonengaho,
Kita, Yoko. 2008. *Utsu to tsukiau* (*Getting Along with Depression*). Tokyo: Seiwa Shoten,
Kusumi, Masayuki, and Jiro Taniguchi. 2020. *Kodoku no gurume* (*Solitary Gourmet*). *Volume 1*. Tokyo: Fusosha.
Miyajima, Kenya. 2010. *Jibun no "utsu" wo naoshita seishinkai no hōhō: Sinshin tomoni genki wo torimodosu tame ni*. Kawadeshobo-shinsho: Tokyo.
Money Plus. 2017. https://media.moneyforward.com/articles/582 (retrieved on July 2, 2021)
Moritaka, Yūji. (Illustrated by Adachi, Keiji). 2010–2014. *Gurazeni (Money Diamond)*. Tokyo: Kodansha.
Nagano, Haru. 2019. *Fuan-san to watashi* (*Fuan-san and Me*). Tokyo: Yamabukishoten.
Netolaboenta. 2016. https://nlab.itmedia.co.jp/nl/articles/1603/01/news048.html (retrieved on July 17, 2020).
Nishiura, Koji. 2019. https://kagiroi.com/publishing-times/7599 (retrieved on June 20, 2021).
Otsuka, Eiji. 2017. *Manga de wakaru manga no rekishi* (*The History of Manga Told in Manga*.) Tokyo: KADOKAWA.
President Online. 2019. https://president.jp/articles/-/28609 (retrieved on June 22, 2021).
Sugiura, Kanna. 2017. "The World's, Japan's and Our Depression (*Sekai to nihon to jubun no utsubyō*)." Japan WHO Association Forum Lecture. https://japan-who.or.jp/wp-content/themes/rewho/img/PDF/library/061/book6402.pdf (retrieved on July 9, 2020).
Takemiya, Keiko and Uchida, Tatsuru. 2014. *Take to Tatsuru no manga bunkaron* (*Take and Tatsuru's manga culture theory*). Tokyo: Shogakukan Shinsho.

Tanaka, Keiichi. 1986. *Dr. Chichibuyama*. Tokyo: Koikeshoin.

———. 2002. *Shinbatsu* (*Divine Punishment*). Tokyo: East Press.

———. 2017. *Pen to hashi* (*The Pen and Chopsticks*). Tokyo: Shogakukan.

———. 2019. *Wakage no itari* (*Youth is not Wasted on the Young*). Tokyo: KADOKAWA.

Tezuka, Osamu. 2010. *Mitsume ga tōru* (*The Three-Eyed One*). *Volume 1*. Tokyo: Kodansha.

Ueno, Kentaro. 2011. *Sayonara mo iwazuni* (Without Saying a Goodbye). Tokyo: KADOKAWA.

CHAPTER 8

Hosokawa Tenten's *Tsure utsu* Series (2006–2013): A Couple's Lived Experience of Depression

Introduction

The previous chapter, "Tanaka Keiichi's *Utsunuke* (2017): An Ode to Depression Tōjisha" showcased Tanaka Keiichi's *Utsu Nuke* (2017), in which the manga artist tells of his experience with depression from the tōjisha perspective with both authenticity and agency. In contrast, Hosokawa Tenten's story about depression is narrated by the wife of a tōjisha. In this chapter, we will examine how the experience is perceived from a family tōjisha's point of view and what advantages this particular perspective has to offer. Throughout the interview, Hosokawa used both *utsu* (depression) and *utsubyō* (depression illness) interchangeably, while referring to the condition as *byōki* (illness) in some instances. Therefore, I will also use the words "depression" and "illness" especially when quoting her.

Hosokawa Tenten's (Fig. 8.1) essay manga, *Tsure ga utsu ni narimashite*, is translated at WorldCat (https://www.worldcat.org/) as "*My SO Has Got Depression*," where SO stands for "significant other." The first volume came out in 2006 at a time when depression was largely stigmatized and misunderstood. Published by Gentosha, the story features her husband's depression and her direct experience as the wife of a depression tōjisha. The manga became a bestseller that year and was followed by sequels, *My Significant Other Has Got Depression Afterwards* (*Sonogo no tsure ga utsu ni narimashite*) in 2009 and *My Significant Other Has Got*

Y. Okuyama, *Tōjisha Manga*,
https://doi.org/10.1007/978-3-031-00840-5_8

Fig. 8.1 Image by Hosokawa Tenten. (Courtesy of Gentosha)

Depression After Seven Years (*Nananenme no tsure ga utsu ni narimashite*) in 2013. The first volume depicts how her husband reacts to developing depression as well as how she manages to survive this most challenging time with him. The second volume tells the reader what happens after the first volume was published while the third volume reflects upon their past seven years since the onset of his depression. When discussing these three related titles, I will refer to them as the *Tsure utsu* series hereafter.

The story of *Tsure ustu* is practically born from the diary kept by her husband, Tsure. In the interview, Hosokawa told me that she had asked him to keep a diary during the course of his illness. She illustrated that episode in another manga, in which she had her character Ten-san say, "Will you keep all of those in a journal because I want to make a book out of it one day" (*Iguana's Bride*, 2015, p. 177). In spite of his strong trust in her, he once rescinded his agreement that she could publish her manga based on his experience. Specifically, he said to her, "No, I take it back. Don't publish it [my story]!" Hosokawa said to me jokingly, "Yes, he resisted it with tears," but added that later he was glad he had agreed eventually and even felt relieved. She said with a chuckle, "Well, he allowed me to read that diary on the premise that his experience would help others. But he never let me read the current one he keeps."

Out of the few manga publishing companies she visited, Gentosha was the publisher that accepted her proposal and eventually published *Tsure ustu*. However, this company primarily publishes novels and technical books. Before she brought her draft to their office, they had rarely published manga. That made me wonder why a non-manga publisher gambled by accepting a story of a "negative subject matter," written by an obscure manga artist. It turned out that the publisher made a commercially clever choice since the manga soon became a bestseller. Although Hosokawa herself could not answer why Gentosha gambled with her work, she speculated that three key factors enabled the unexpected popularity of *Tsure ustu*, which are later provided in the section "Reception and Generalizability."

The Title and Target Readership

The original Japanese title of the first volume (published in 2006) is ツレがうつになりまして (*Tsure ga utsu ni narimashite*). The 2009 and 2013 sequels have the same phrase, *tsure ga utsu ni narimashite*, as part of their titles. The word *utsu* is short for *utsubyō*, referring to clinical depression.

Another expression in the title that tells of the author is *tsure* (significant other). Hosokawa chose the word over more common terms addressing a male spouse, *shujin* and *danna* because "I regard him as my *pātonā* (partner). Both *shujin* and *danna* are terms indicating power structure. Our relationship is nothing like that." In Japanese, the most common reference to a male spouse is *shujin*; however, as shown with its *kanji* characters 主人("master"), its implication is a person whom you serve and is used in phrases such as *ikka no shujin*, ("head of the household"). Originating from the Sanskrit word *dāna* ("person who provides you with alms"), the word *danna* similarly connotes that the referent is the "provider" in a relationship. Thus, she dislikes *shujin* because "he is not my 'master.'" Nor does she accept *danna* because "the word implies the pecking order [between husband and wife]." In comparison, *tsure* (a casual term of "husband," close to the English equivalent of "significant other") and *otto* (a formal term of "husband") are neutral.

Hosokawa told me that the title was taken straight from what she repeatedly said to editors at several publishers after disclosing her husband's illness in order to convey her urgency at finding more work as the sole breadwinner.[1] Once her husband was diagnosed with clinical depression, he left his employment voluntarily to restore his health. Although Hosokawa had already made her debut as a manga artist, she was a rookie and had trouble finding work in the industry. Undeterred, she continued to visit other publishers to desperately look for more work. "So, I told people something like 'My significant other came down with depression, so please give me a chance to work for you!'" she said laughingly. Eventually, her persistence bore fruit, and she managed to get some comic magazine contracts from several publishers. Then, she decided to write a comic about her husband's depression. Finding a publisher for that project was more challenging than she had thought. "Some editors probably were prejudiced and responded, 'Such a negative subject matter doesn't work for us,'" she added. Fortunately, one of the publishers accepted her proposal for *Tsure utsu*. "In the end, this manga was published as a book [a *takōbon* version of her manga]. I'm glad I was undeterred, trying one publisher after another."

The first title of the series led to Hosokawa's breakout from her then-stagnant manga career. Commenting on *Tsure utsu*, the abbreviated version of her three related titles used by fans and the media, she stated, "Only after the book [*takōbon*] was released and became a hit, was it referred to as *Tsure utsu*. Then, I thought, 'Oh, that's how it works.'" It was only later that she learned about the rumor of an 'enigmatic four-syllable word' (*nazo no yonmoji*) in the manga industry but had no knowledge of it when she created the title. (The coinage and meaning of the *nazo no yonmoji* is explained in more detail in chapter "Tanaka Keiichi's *Utsunuke* (2017): An Ode to Depression Tōjisha".) While interviewing her, I sensed her genuine surprise about the unexpected success of *Tsure utsu* as she sounded like a star-struck Cinderella. The section of "Reception and Generalizability" provides the social backdrop of her Cinderella story, drawing from Hosokawa's own description of how a low-status manga artist unexpectedly achieved instant success.

As she said during our interview, as well as in another interview (*Josei no Hiroba* 2017), her purposes of publishing *Tsure utsu* were to correct misunderstandings of depression (e.g., personality-induced illness) and normalize the condition ("I wanted people to understand it can happen to anyone"). For this reason, the primary readership Hosokawa had in mind were lay Japanese, particularly families whose members do not know much about depression. Using the term *comic essay*, one of the variants of essay manga, the author defines *Tsure utsu* as a "comic essay that depicts an illness" written with two intertwined purposes. One purpose was to inform the reader of how it feels to be depressed through Tsure's experience and how the illness impacts everyone in the household based on the author's experience. The other purpose, which I sensed during the interview was her primary mission, was to correct misconceptions about depression that were still pervasive in society.

Some of the misconceptions that she encountered during Tsure's illness are that a "gloomy personality causes depression (*seikaku ga kurai kara naru*)," "people who are afflicted with depression tend to be mentally weak (*utsubyō ni naru nowa ki no yowai hito*)," and "a person with a negative personality gets depression (*seikakuga ushiromukina hito ga utsubyō ni naru*)." These comments reflect the old myth of *seikaku-in* (personality cause) discussed in Kitanaka Junko's (2012) seminal book on depression in Japan, *Depression in Japan*. Prior to that personality theory, Shimoda Mitsuzo, a prominent psychiatrist in the 1950s, argued that depression is a "'biological response for self-preservation' that occurs at the height of one's fatigue—a protective mechanism of adaptation" (1950: 194, cited by Kitanaka 2012, p. 71). Stressing that depression is more directly associated with stress and other factors than with a certain personality, Hosokawa raised her concern about common beliefs held by the public by publishing this manga. When she told various people that Tsure had been diagnosed with depression, one common reaction was "Oh, I didn't know he was such a gloomy person." Although she attempted to say, "No, it's not so," each time, they seemed to be stuck with that stereotype. Throughout the interview, she asserted, "It has nothing to do with one's personality."

This popular myth about depression seems to persist among people, including tōjisha themselves. In the previous chapter, I presented manga artist Tanaka's comment that people with certain personality types are more susceptible to developing depression (*Gendai Business* 2018). In that interview, Tanaka's peer artist responded, "People who tend to work toward a goal seriously are more prone to it [depression]. I myself am one of them, setting up an even higher goal once the first one is achieved." It is important to clarify that according to the *DSM-5*, major depressive disorder is a condition that meets five or more symptoms such as prolonged depressed mood and reduced interest in activities that the person previously considered pleasurable (2013: 160). What causes depression is still unknown, although a host of factors from genetics to environmental stress to personality have been suggested as contributing factors.

Hosokawa explained that many misconceptions were circulating because many people still did not understand what depression was. Her assertion is convincing when we reflect on the history of cancer, especially how the illness was misconstrued as "contagious" when medical research was limited and the media did not even mention the word *cancer* openly in the early 1990s. She emphasized the ubiquitous nature of depression, regardless of personality, by saying, "This is the type of illness that anyone can be afflicted with. People should know that fact, especially if someone in their family becomes depressed in the future. Personality is beside the point. I wanted to send that message to the public." As Tsure himself describes in an essay inserted in *Tsure utsu* (Vol. 3, p. 44), he was an optimist prior to the onset of depression. As Hosokawa depicts in one episode (Vol. 1, pp. 18–19), Tsure used to be the one who possessed emotional strength, while Hosokawa's character, Ten-san, who was a pessimist by nature, constantly needed Tsure's encouragement, particularly when mulling over her stagnant manga career. Therefore, it was all the more shocking to Hosokawa to witness how her husband's demeanor had changed during a year of suffering from depression.

Unlike the first volume, Hosokawa produced two volumes as sequels mainly to tell her *Tsure utsu* fans how Ten-san and Tsure have managed several years of his recovery as a married couple. Because she never imagined the huge success of the first volume, she had not even thought about writing follow-up stories. "At first, I was a bit flabbergasted when I was asked to make a sequel," she said with a chuckle. "Definitely, I recommend the first one, which illustrates what actually happened. Especially those who are ill and their family members will find useful information about the condition," she added. However, in my view, the sequels offer family members of the depressed something the first volume does not: pragmatic information. For example, the second volume explains technical terms such as the Services and Supports for Persons with Disabilities Act (*shōgaisha jiritsu shienhō*) and cognitive behavioral therapy (*ninchi ryōhō*) and key notions including one's relationship with a municipal office. She told me that her respective editors (each volume had a different editor) asked her to include more technical information; consequently, the sequels have slightly more text than drawings. The third volume is equally enjoyable in that it tells the reader about Tsure's recovery processes, the background of the manga's television and film adaptations, and the birth of their baby boy.

Although the tōjisha may also pick up this manga, Hosokawa told me that it might not be wise to recommend it to people who are in the midst of a deep depression. She pointed out that they may find reading itself difficult (as it happened to Tsure). "So, I'd rather recommend it to them after they get over the hump," she clarified. On reflection, the author was probably hinting at the condition of anhedonia, reduced motivation to experience pleasure, affecting activities that the person enjoyed previously such as reading and listening to music. Thus, depressed individuals especially may not desire to read about the very cause of their suffering, according to her.

The Author's Background

Drawing has been Hosokawa's favorite pastime since childhood (Vol. 3, p. 82), and her dream of becoming a *mangaka* came true at age twenty-seven (*Iguana's Bride*, p. 49) through a monthly *shōjo* manga magazine named *Bouquet DX* in 1996 (*Iguana's Bride*, p. 200; Mochizuki 2010, p. 179). For the next ten years, until *Tsure utsu* became a hit in 2006, Hosokawa was a struggling comic artist. Her debut work was a sixteen-page short story titled, *Sora no sakana* (*A Fish in the Sky*) (Mochizuki, p. 179). *Bouquet DX* discontinued her second comic's serialization. It lost favor as her drawing style leaned toward *Garo*,[2] an artistic style that would be too daring for the *shōjo* manga genre (Mochizuki 2010, p. 181). At that time, Hosokawa was a big fan of Tsuge Yoshiharu, one of *Garo*'s regular manga contributors (Mochizuki 2010, p. 182). She then realized her true passion was not to write romantic stories for *shōjo* manga (*Iguana's Bride*, pp. 98–99).

In our interview, she identified Takano Fumiko as her favorite, most admired manga artist. Hosokawa's pen name is Tenten, a duplicate of *ten* (ferret) coined by her husband, Tsure. The animal happened to be one of the animals that appeared in Takano's work, so Hosokawa liked that pen name because of its association with this beloved *mangaka*. "It's a weasel-like animal from which we get fur. He [Tsure] said he had picked it because he thought I looked like *ten*," she explained. Hosokawa was selected for the last name after the then-Prime Minister of Japan, Hosokawa Morihiro (Mochizuki 2010, p. 153). While she was struggling professionally, she experienced mild depressive symptoms such as breathing difficulty, loss of energy, oversleeping, feelings of unworthiness, and negative self-talk, which are characteristic of work-related grief but did not induce clinical depression in her case (Mochizuki 2010, p. 182). In the meantime, Tsure's career steadily improved, and he eventually secured a high-paying job at an IT company at the peak of the dot-com bubble (1999–2000).

Hosokawa was blessed with a good working relationship and emotional support from a *tantō henshūsha* (manga editor) as she published each volume of the *Tsure utsu* series (Vol. 3, p. 129). In particular, Sugano Yumi, the editor of the first volume, was very eager to have this manga published, since no manga about depression had been written before from both the patient and family's perspectives (Vol. 2, p. 27). Hosokawa told me that Sugano allowed her to write the story "just the way I wanted." Although she never asked Sugano for the reason, she took it as an indication that her editor respected her artistic and storytelling sense. Hosokawa came to the interview without her editor. Unlike my Kodansha-mediated interviews, Hosokawa's publisher, Gentosha, arranged this interview without the accompaniment of any of her three editors. Thus, I had no way of verifying Hosokawa's assumption. Another interpretation might be that the editor valued the unedited remarks and frank opinions of the patient and family.

The second and third volumes of the series depict the days in which Hosokawa becomes a popular *mangaka* and their only child is born. As a *sengyō shufu* (full-time male homemaker), Tsure assumes the primary role of parenting and household chores. During the interview, Hosokawa commented, "Our son is in the fifth grade. Now, we take care of him with fifty-fifty parental responsibilities. But when he was a baby, Tsure did everything. I was able to concentrate on my work, but he started scolding me [laugh] saying, 'You take no part in parenting!' and 'You don't care for your family (*katei wo kaeriminai*)!' So, we now negotiate our [parental] tasks as we go." In a Japanese household, *katei wo kaeriminai* is typically a female spouse's complaint against her husband, accusing him of hardly ever prioritizing his family. The other way around is an indication of the role reversal. That explains why Hosokawa, now age forty-two, has a strong aversion to the connotation of *shujin* as "master," as explained earlier in this chapter, and prefers calling her husband *tsure* ("significant other").

Hosokawa is a family tōjisha. As mentioned earlier, the *Tsure utsu* series is the first manga about depression written from the perspective of a family *tōjisha*. Hosokawa described how painful her family *tōjisha* experience was, especially during the peak of Tsure's illness, as follows:

> I felt really helpless. Even if I tried to lend a helping hand, it was mostly ineffective. The best I could do was only to watch over him earnestly (*hitasura mimamoru shika dekinai node*). To someone who watched over him, it was unbearable. Even if that was his illness, I felt I was being abused by the situation (*furimawasareteru tte kanjide*). Not a single day went by without asking myself, 'What did I do to deserve this?'

Of the several manga artists I interviewed for this book, Hosokawa is an exception in that she went through this sort of retraumatization. Some artists told me that it was almost therapeutic to recall and retell the experience after their recovery. Others like Okita (chapter "Okita Bakka's *Gaki-Tame* Series (2011–2013): A Memoir of a "Troublemaker" Aspie Girl") found that the task of converting the experience to a graphic memoir gave them an objective and coherent perspective of their tōjisha experiences from initially fragmented pieces of memory. Komame Daruma, the manga artist mentioned in chapter "Essay Manga", ended up finding the strength and boldness to fight the illness while also keeping a psychological distance from the experience. In the interview, Hosokawa pointed out, "I think Tsure was able to objectify his lived experience by writing about it. I still find it difficult to objectify or transform my experience, though." However, she also stressed that she had benefited from the experience because she came to learn how those who live with the patient can fall victim to the circumstances when it comes to mental health issues. She understands how torturous it can be to get stuck in that situation.

The stress of family *tōjisha* may be enormous for the spouse if the household consists of only a wife and husband. A case of *tomodaore* (falling together) may ensue, especially when the healthy spouse neglects self-care. When Hosokawa gives a talk, she makes sure she talks about the spouse's well-being. "I tell the audience, 'If you are not well, you can't take care of the loved one well.' Your health is your priority. So many people just think about the sick one and say, 'I don't know what else to do!' So, I tell them, 'Please remember to take good care of yourself.'" Hosokawa has heard of many cases where both husband and wife became depressed; in some cases, the couple ended up divorcing as the situation deteriorated. "Of course, depression isn't contagious. But if you are deeply involved with the illness [as a caregiver], chances are you might develop it yourself. I know for a fact that the negative energy that the patient generates can be so powerful that you can get depressed merely by being together," she added.

Therefore, Hosokawa shares with other family *tōjisha* the invaluable lesson that she learned: keep a healthy distance from each other and protect yourself by creating your safe zone (*jibun no kūkan*). "That is because even the unafflicted one inevitably becomes afflicted if the couple shares the same space twenty-four seven," she warned in the interview. She revealed her method of "separation" through the use of *fusuma* (Japanese-style sliding partitions/panels). She said that she was able to survive the testing time by sleeping and working in an area partitioned by the *fusuma* in the house. "I didn't see him but could come immediately if he needed me."

Besides keeping a safe distance and maintaining self-care, Hosokawa stresses the value of fostering open communication within the family. "What helped us as a couple during that tough time was talking over everything. To understand each other, you have to express your thoughts to the other with words. If we're just silent, we don't know what the other is thinking. So, we keep this motto in mind: 'Let's have a talk, no matter how insignificant the matter is (*kudaranai koto demo iikara nandemo hanasu tteiuno wa kokoro gaketemasu*),'" she told me. From my own experience as a daughter of a mentally ill parent, I could relate to what she meant by the "negative energy" and how "contagious" it might be. The importance of self-care as well as support at the medical and community levels, cannot be overemphasized for all family *tōjisha* of the ill, including parents, children, and siblings. "Mental illness affects the entire family indeed," she chimed in.

Although the story is narrated by her character, Ten-san, Hosokawa's *Tsure utsu* series sheds light on *tōjisha mesen* (tōjisha perspective). Chapter "Tōjisha" highlights the significance of listening to *tōjisha* voices in association with the disability rights motto, *Nothing about us without us.* As stated there, disability rights activists such as Fujii (2014/2016) have argued in their disability memoirs in which the *tōjisha* share their lived experience of mental distress that they can help increase public awareness of such "invisible" disabilities as depression. In the interview, Hosokawa explained two methods of incorporating *tōjisha mesen.* First, Hosokawa created Tsure's depression episodes based on a journal that he kept during that time. She stressed that the contents of the character Tsure's speech and inner voice were derived from that diary. "In that regard, it [this manga] contains plenty of *tōjisha mesen,*" she asserted.

Hosokawa's second method of including *tōjisha mesen* is making space for the tōjisha's real voice in the manga: each volume of the *Tsure utsu* series has "*Tsure no tsubuyaki* (*Tsure's whisper*)," a one-page essay written by Tsure himself inserted between episodes. "My [first volume's] editor let me take full control over the content. But after I had finished the manuscript, she made one request, 'I want to include his own account of it [depression]. Can he write something as well?'" She added, "Looking back, I'm glad that they were included. Having both my [family] perspective and his [*tōjisha*] perspective made the story more realistic (*riariti ga mashita*), I think." Thus, the unique characteristic of this manga is that, while the story is narrated by the wife, the husband's experience of depression is represented with the *tōjisha mesen.* My claim that *Tsure utsu* is "a tōjisha manga" is further substantiated by her conscious inclusion of the *tōjisha mesen* through these two approaches.

In the interview, I pressed on by asking why she did not opt to have Tsure narrate the whole story if *tōjisha mesen* is of primary importance to her as the author. She answered laughing, "If this story were told only with *tōjisha mesen*, it would be *uttōshii* (depressing)." She asked me to imagine a manga where the narrator mumbles nothing but "I can't do this" or "I can't do that" endlessly. Her response is reasonable, considering that the manga's serialization commenced when Tsure was still too ill to be actively involved in the project. "It'd be grueling for ordinary people to keep on reading such a story [told from a depressed person]. On top of that, the medical treatment does not immediately result in a dramatic happy ending (*gekitekina happīendo*)," she said. Then, she shared her insight into storytelling: the voice of the family *tōjisha* creates a buffer around the otherwise depressive narrative of illness. She thought that the able-bodied narrator might say something like "Let's stay positive together as a family" and end the story with hope for the family. Arguably, Hosokawa's family discourse is just one of many effective tactics, and the authors of other, similar graphic memoirs on disability and illness would probably argue otherwise. For instance, in his *Shissō Nikki* (Disappearance Diary) and *Utsu utsu Hideo nikki* (Hideo's Depression Diary), Azuma Hideo capitalizes upon *gag* manga's slapstick humor to lighten up the otherwise gloomy episodes of his mental health struggles. Likewise, Okita Bakka (chapter "Okita Bakka's *Gaki-Tame* Series (2011–2013): A Memoir of a "Troublemaker" Aspie Girl") has the comedic character, Bakka-chan, narrate the story including serious scenes of bullying, sexual harassment, and suicide attempts, all without graphic realism. In *Utsu nuke*, Tanaka Keiichi (chapter "Tanaka Keiichi's *Utsunuke* (2017): An Ode to Depression Tōjisha") tells his lived experience of depression by sharing it with a sidekick character, Kaneko, in the two-person standup comedy style, *tsukkomi* and *boke* (described in chapter "Essay Manga"). Kaneko's presence prevents Tanaka's narration from being too depressing for the reader. Whatever tactic is selected, the authors of *tōjisha manga* appear hypersensitive to the emotional impact of their narrative on the reader.

Hosokawa's Drawing Styles and Storytelling Characteristics

For manga artists, it is a challenge to figure out how to visually represent such invisible matters as psychological symptoms and the character's emotions and inner thoughts. Some artists employ conventional signs of manga semiotics (*manga no kigō*) while others invent their own unique imagery. Earlier in the interview, Hosokawa described her process as something natural, saying, "I don't consciously think about that when I'm drawing. It happens unconsciously." I pressed on by presenting the following specific images from the *Tsure utsu* series that are representative of her drawing, asking her to elaborate how she came up with them.

Characters of Ten-san and Tsure

In *Tsure utsu*, Hosokawa's character, Ten-san, appears as a round-faced woman with a double chignon hairstyle (e.g., the top panel of p. 18 in

Fig. 8.2). I asked her if she had had that hairstyle in the past, since she had short hair at the time of our interview. She replied, "No, no. When my penname Hosokawa Tenten was decided, 'Tenten' sounded like a Chinese word to me. That's why I drew my character with chinoiserie." (To clarify, ferrets inhabit Japan and South Korea but not China. It is possible that she associated Tenten with the typical panda bear names, "Shanshan" and "Shinshin.") Ten-san is often drawn with her face overblown so that it fills almost half the frame as in the first panel on page 18 and the second top and the first middle panels on page 19 in Vol. 1 (Fig. 8.2). Regarding the implication of her enlarged face, Hosokawa explained that it is her way of emphasizing the character's inner thoughts. For example, her impactful emotional state of being stressed is expressed with sweat beads on her enlarged face (top panel on p. 18). When she screams "Why?" with anger, her face is also drawn as disproportionately large (top, second panel on p. 19). The same trick is used in depicting her frustration with her utterance, "What the heck's depression?! (*Daitai utsubyō tte nani*?!)" (middle,

Fig. 8.2 Image by Hosokawa Tenten. (Courtesy of Gentosha)

Fig. 8.3 Image by Hosokawa Tenten. (Courtesy of Gentosha)

right-side panel on p. 19). She told me that she finds it rather difficult to draw emotionless characters or a peaceful state of mind.

Hosokawa met Tsure for the first time when they were students at Setsu Mode Seminar, an art vocational school in Tokyo (*Iguana's Bride*, p. 27), which was founded by the famous Japanese artist Nagasawa Setsu and was closed in 2017. Unlike Hosokawa, Tsure did not pursue his career in art because he decided that he "was not artistically gifted" (Mochizuki 2010: 125). Born in 1994, the year of the first Tokyo Olympics (thus six years her senior), he was an optimist by nature and loved classical music and cooking among his many other hobbies (most of which he would lose interest in when he became severely depressed). At first, he was a "super salaryman" (Vol. 1, p. 4) happily committed to his work at the IT company as shown (see right panel of Fig. 8.3). In 2002, he felt extremely energetic both in personal and professional spheres. Even after the company held a massive layoff and the volume of his workload increased exponentially in 2003, he was able to manage his productivity.

However, his health began to show warning signs, first with severe anemia followed by sleeplessness and alternating diarrhea and constipation. Three months later, in 2004, Tsure was diagnosed with depression at age 39 (*Iguana's Bride*, pp. 200–201; Mochizuki 2010, p. 19). A dramatic change in the state of his well-being is observable on pages 4–5 of Fig. 8.3. His insomnia led to an inability to concentrate and severe headaches in the daytime, which he believes signaled the onset of his depression. By the beginning of 2004, he was getting the blues, unable to focus on the task at hand, making numerous mistakes, and even experiencing auditory hallucinations on occasion (Mochizuki 2010, pp. 28–29). Unlike normal feelings of sadness and grief, clinical depression is a serious condition accompanied by five or more symptoms such as persistent depressed mood, loss of energy, markedly diminished interest in previously enjoyed activities, feelings of worthlessness, inability to think or concentrate, and insomnia or hypersomnia (DSM-5 pp. 160–161). During his severe depression, Tsure's specific symptoms included the sudden loss of speech, crying bursts, and the inability to count money or ride on a crowded train. Six months later, at the beginning of 2004, he was finally diagnosed with depression and prescribed medicine (Luvox, a selective serotonin reuptake inhibitor, or SSRI, a common type of antidepressant in Japan). He continued to suffer from these symptoms for three more years from 2003 through 2005 (Vol. 1, pp. 18–19; *Iguana's Bride* pp. 202–203).

As another common symptom, the depressed may experience recurrent thoughts of death (*DSM-5* p. 161), and the risk of suicide is quite high in this population (p. 167). One day, Tsure confessed to Ten-san, "I feel like killing myself," and although speechless at first, she responded, half-jokingly, "You're just kidding!" (Vol. 1, p. 60). However, Tsure continued to have recurrent suicidal ideation. When Ten-san is emotionally vulnerable one day and her patience is diminished, she inadvertently makes a series of hurtful comments to Tsure during their argument and issues an ultimatum: "If you keep saying, 'I'm useless,' go admit yourself to the hospital, and don't come back until you get better!" As depicted on page 61 of Volume 1, Tsure alleges in his diary that Hosokawa repeated the same verbatim eight times during that incident. Their verbal fights and reactions escalate, and Tsure throws a heavy footstool of 15 kilograms or approximately 33 pounds at her (second, left panel on p. 62 in Fig. 8.4). Ten-san snaps and screams at him, "You murderer! You could have crushed my head with this [heavy footstool]!" (third panel on p. 62). In response, Tsure shrieks and runs to the bathroom. Unbeknownst to Ten-san, he

Fig. 8.4 Image by Hosokawa Tenten. (Courtesy of Gentosha)

then tries to choke himself in there, only to come to his senses and stop midway (from top, left panel to three middle panels on p. 63 in Fig. 8.4).

In these panels, Hosokawa utilizes multiple *manpu* signs.[3] For example, Ten-san's face is drawn with a wedge-like mark on her temple (see third panel on p. 62 and last, right panel on p. 63), signifying that she is upset or having a temper tantrum. In Japan's classic theater performance, *kabuki*, a similar "anger" sign called *kanshaku-suji* (anger marker) is painted on the actor's face to visually symbolize that emotion. Another classic example of *manpu* used in this sequence of panels is a sweat bead on her face (see second, left panel on p. 62). In manga semiotics, the drawing of sweat signifies a feeling of being fretful, anxious, ashamed, or a similar state of emotion of the character. This *manpu* sign is derived from a Japanese phrase, *hiya ase wo kaku* (break into a cold sweat in an anxiety-provoking situation). Rather than representing physical sweat drops, the sign signifies an emotional state of the character. More examples of typically used *manpu* signs, including a "happy" note (bottom right) implying the character is in a good mood or completing a task with ease, can be found in Kōno Fumiyo's *Gigatown: Manpu zufu* (2018), a pictionary (illustration-based dictionary) of *manpu* signs.

In this manga, Tsure's character is drawn with an onion-shaped head as shown in Fig. 8.5. In the interview, Hosokawa told me that it was drawn "just as he looked back then." Comically, a pair of insect antennae are growing out of his onion head. The antennae are said to "pick up various hardships" in the air (p. 5 of Vol. 1) and receive "signals of depression" (p. 7 of Vol. 1). As the bottom left image on page 8 (Fig. 8.5) shows, Hosokawa artistically expresses the presence of Tsure's depression with a dark shadow around his head to intensify the expression of his depressed mood. She describes how she came up with such visual symbolism as follows:

Fig. 8.5 Image by Hosokawa Tenten. (Courtesy of Gentosha)

> "Tsure changed 180 degrees from what he was like before depression—his remarks, behaviors, and so on. To metaphorically express that transformation, I drew the antennae, thinking, 'His words and deeds can be accounted for if he has antennae in his head and is sensing things that I cannot pick up. These antennae are invisible but do exist.' That's how I reasoned with myself as I drew it." Once Tsure's depression goes into remission (*kankai*), the character loses the antennae on his head and the dark shadow on his back (see p. 9 of Figure 6.5).

Representations of "Getting the Blues"

Hosokawa first represented the illness of depression with a Halloween, ghostlike figure with the label "*utsubyō*" on its back (Vol. 1, p. 8). "My first impression was that Tsure was possessed by something incomprehensible. His mood swings were so severe that he would suddenly feel low or become unable to get out of bed. [The ghost is] probably a reflection of that mental image I had." However, after the first episode, she abandoned the figure and began to use two primary drawing gimmicks to illustrate the character getting the blues. One way to imply a depressive mood is, as mentioned earlier, by drawing the dark shadow around the character (see bottom right panel in Fig. 8.6). Hosokawa explains, "Depressed people appear to me as if they are carrying shadows (*kuroi kage wo seotteiru yona kanji*)." Also shown in the same right-side bottom panel of Fig. 8.6 are vortex-like, coiled air puffs floating nearby (often accompanied with a *manpu* sign, *doyōn*, short for *donyori*, "of dull, gloomy atmosphere"). Drawn with his shadowed face, this signifier of the coiled puffs expresses Tsure's hopeless outlook. Applied frequently throughout the *Tsure utsu* series, this image is also seen in Fig. 8.5. In the manga, she calls these coiled air puffs "negative aura" (*mainasu ōrora*) (Vol. 2, p. 117). In the interview, she explained that they symbolize the person's intense feelings of a sharp emotional downturn, including depression.

These symbols not only express Tsure's depression episodes but also show the Ten-san character feeling the blues. Figure 8.7 shows an image of Ten-san, sweating and groaning while working on her manga, accompanied by her inner voice in the white field, which reads: "Writing a story about depression was much more painful work than I had thought it would be" (Vol. 2, p. 31). In the interview, Hosokawa confided in me, "The emotional pain was so punishing at that time that I still find it too traumatizing to watch the drama and film [adaptations of *Tsure utsu*]. I

Fig. 8.6 Image by Hosokawa Tenten. (Courtesy of Gentosha)

Fig. 8.7 Image by Hosokawa Tenten. (Courtesy of Gentosha)

went through hell, so writing the manga was like reliving that experience. I was writing, then crying, writing again, and then crying. It was a toilsome undertaking, indeed." She also explained that she wept a lot while working, not only when she immersed herself in the past, remembering some painful episodes, but also when she paused to return to the present, appreciating that Tsure is well now.

As Hosokawa's drawing strategy, the visual symbols of air puffs and a shadow, as well as the word "*doyōn,*" serve as metaphors that mean "going pale," "turning blue," or "being depressed." In *Gigatown*, Kōno (2018) describes the manga semiotic sign of coiled puffs as a symbol that represents a dark psychological state. Furthermore, Kōno explains, shadow lines are often drawn on the character's forehead or below the eyes as a visual metaphor for a mental state of turning blue (*aozameta jōtai*), accompanied by an *onyu* sign (sound effect) expressed with the word "*saato*" (quickly) implying that the person is "going pale" (p. 24). Although these are rather conventional signs of manga semiotics, Hosokawa probably invented a version that is uniquely hers. To emphasize the psychological torment that the character is experiencing, she draws not only the puffs but also shadow lines around his/her temples. These signifiers are seen, for example, in the scenes of Tsure in a cold sweat and feeling overwhelmed with the sardine-packed space of the morning train (*asa no manin densha*, Vol. 1, p. 26), his trying to fight fatigue and feeling numb (Vol. 1, p. 30), Ten-san's getting disappointed with *gakkari* ("Shucks!") (Vol. 2, p. 97), and her worrying excessively with *fuan* ("worries") (Vol. 3, p. 89).

*Depression Waves (*Utsu no Nami*)*

To express Tsure's mood swings, Hosokawa uses a metaphor of "waves of emotion." As shown in the middle panel of page 44 in Fig. 8.8, Tsure is about to be swallowed up by a big wave of melancholy (*yūutsu no nami*). "As I got a close look at his condition, I noticed the pattern of his depression. At first, his emotional dip appeared to happen suddenly and hit us off guard. But actually, his deep depressive mood was preceded by small mood changes. So, the cycle was like waves—a succession of small waves leading to a big surf. The more I saw the pattern, the better I could anticipate the big wave coming. And I was able to warn myself, thinking, 'Oh, a big wave will come soon!'"

This association of waves and emotions is not a unique strategy created by Hosokawa, however. In Japanese, emotions are commonly referred to

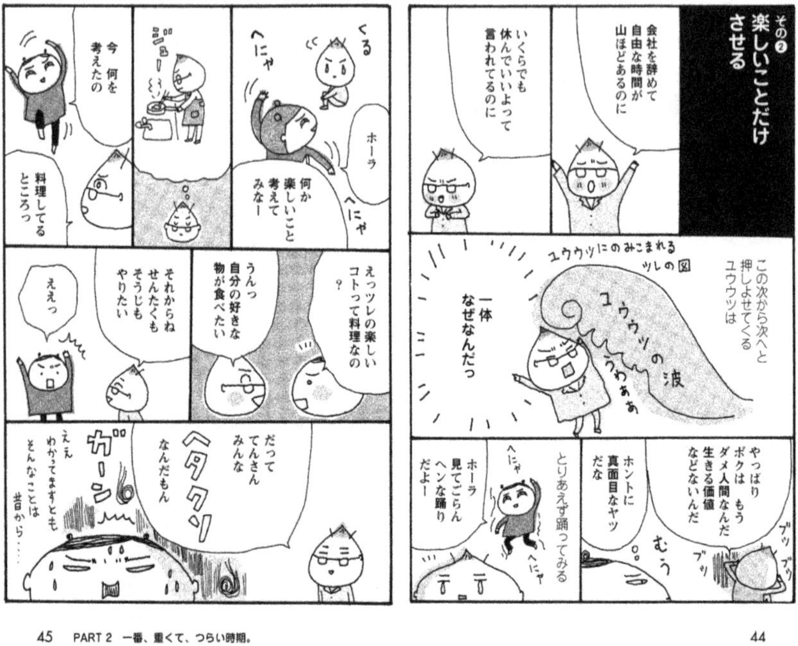

Fig. 8.8 Image by Hosokawa Tenten. (Courtesy of Gentosha)

as "waves," and in some psychology literature, mood disorders such as clinical depression are said to come to the shore repeatedly like waves. Although the pattern of recurrence may seem unpredictable at first, one can observe it and learn to avoid getting knocked down by the wave. As I analyze various manga authors' images, I have come to notice that many manga images are actually derived from the semantic association with Japanese phrases; for example, *hiyaase wo kaku* (break into a cold sweat) is an expression that implies a mental state of fear or nervousness and is typically represented in manga with sweat beads on the character's face. It appears Hosokawa applies this process of inventing visual expressions via semantic association with existing linguistic expressions such as "waves of emotions."

Turtle Bed (Kame Futon)

As with the top image on page 8 in Fig. 8.5, Tsure is drawn curled up in bed and crying tears of depression (with *shiku-shiku*, or "Wah-wah") in a small, hill-like image. This iconic image appears on many pages of the series and is called "*kame futon*" (turtle bed). She explained how she came up with that drawing by saying, "Quite often Tsure would stay in bed and weep with the *futon* cover over him. Back then, we had a pet turtle whose behavior I often observed. [Because of their resemblance] I labeled it [Tsure's behavior] '*kame futon.*'" She added, "The bigger his depression wave, the higher the percentage of his turning into the *kame futon*, I noticed." Unlike its cute and comical representation, the full-blown episode of Tsure's depression was serious. However, the humorous drawing helps lighten up the somber mood of the scene, making it easier for the reader to absorb the information. As Yamashita (2009) emphasizes, Hosokawa's healing character drawings (*datsuryoku kei kyarakutā*) as well as her use of the humor that is characteristic of essay manga are two advantages of this manga in telling a story of depression without depressing the reader.

Hosokawa artistically differentiates this "hitting-bottom" behavior of weeping and hiding out in bed from a milder state of his depression. As shown in the top left and middle panels on page 39 in Fig. 8.6, Tsure is depicted as being wrapped in sleeping bag-like bedding, muttering to himself, "*butsu butsu*" (moaning), when his feelings of worthlessness are not so excessive. "I probably found that behavior interesting enough to create a manga image," she said laughing. To depict Tsure's vacant facial expression (*tamashii ga nuketa kanji*), she employs the conventional sign of "white eyes" as in the right bottom image of Tsure on page 5 in Fig. 8.5. Although this standard manga sign connotes "the experience of a traumatic shock" (Kōno 2018: 49), Hosokawa uses it as an expression of Tsure's lifeless mental state.

Hope and Happiness

To represent hope or a mild state of happiness, she decorates the character's background with starlight and flowers as seen in the bottom left panel on page 90 in Fig. 8.9. "I use those to contrast the scene sharply with opposite emotions of Tsure's ['feeling blue,' for instance] so that the reader can easily tell the difference," she explained. Starlight and flowers

90

Fig. 8.9 Image by Hosokawa Tenten. (Courtesy of Gentosha)

are conventional manga signs that appear to be associated with (and possibly derived from) Japanese idiomatic phrases such as "*hikari kagayaku mirai*" (future sparkled with lights), "*komyō ga sashikomu*" (lights of hope are coming in), and "*hanabanashiki kibō ni michita jinsei*" (life full of blossoming hope). However, these signs are typically used in *shōjo* manga to highlight the heroine's upbeat mood as used with the character of an instructor in the bottom right panel on page 90 [Fig. 8.9]. As a former *shōjo* manga artist, Hosokawa seems to have a variety of such artistic gimmicks handy.

Textual Strategies

Hosokawa also invents words to incite humor; one example is *uchū kaze* (space cold). "It [depression] was a mysterious illness to me. The universe is equally mysterious. I named it so," she told me. In the manga, the character Ten-san thinks to herself, "When one is depressed, mysterious things happen. It [depression] is like a mysterious cold that extraterrestrials catch. Oh, I have an idea: I'll call it a 'space cold'" (Vol. 1, p. 55). His depression is referred to as "*uchū kaze*" again (Vol. 1, p. 59). As also described in chapter "Tanaka Keiichi's *Utsunuke* (2017): An Ode to Depression Tōjisha", depression is frequently equated to *kokoro no kaze*, or "a cold of the heart" in Japanese cultural context (Kitanaka 2012: 14). In order to destigmatize the illness, the word "cold" was elected in the early 2000s, implying that depression was ubiquitous: like a cold, anyone can be afflicted with depression. It was part of a campaign by both psychiatrists and pharmaceutical companies in Japan to pitch imported antidepressants as an effective cure like a cold medicine (p. 79). Since then, the metaphor of *kokoro no kaze* has stayed in the mind of lay Japanese. Hosokawa, however, turned it into a comical image of an alien's cold.

I asked her how she came up with catchy phrases such as "turtle bed" and "space cold." She replied laughingly, "I like to invent words and phrases when the existing vocabulary does not have a better way of expressing things. When I can't understand something, I invent a label to comprehend it myself. It's my characteristic." I pointed out that her manga fans loved those expressions used in the series. She responded, "That's right. It kind of surprised me actually. I had no idea readers would comment on something like that. To me it was nothing special; I made up those phrases and used them matter-of-factly, like 'Hey, you're back in your turtle bed!'" (*mata 'kame futon' ni natteruyo*).

Readability and Other Considerations

Readability is an important aspect of essay manga. For such considerations, Hosokawa told me that one decision she made consciously was to keep her storytelling as comprehensible as possible with simplified drawings and reader-friendly language because she believed that those suffering from depression were the potential readers of *Tsure utsu*. "When they are suffering, they find it hard to read, especially long stretches of text. Tsure himself became unable to read newspapers and novels, things he used to love to read. So, I intentionally made the manga easy to read." As discussed earlier in this chapter, due to anhedonia, the depressed may no longer find reading enjoyable even if they used to love it. For this reason, easy-to-read material is critically important to those experiencing the symptoms of loss of energy and diminished ability to concentrate as they may find manga the most accessible form of getting information on the illness. Another advantage of essay manga is suggested by Hata (2013): the medium is able to not only present comprehensible information with shared manga semiotic signs, cute characters, and exaggerated emotions, but it also makes space for both empathy and criticism on the reader's end. For example, as shown in Fig. 8.6, Ten-san first feels "*uttōshii*" (annoyed and dispirited), hearing Tsure's self-reproach all day long, but soon criticizes herself for acting similarly when her manga career was down in the dumps. According to Hata, because it is uttered by the family tōjisha who narrates the story, her comment is received as the character's speech, rather than the author's self-criticism, which in exchange allows the author more room to further reflect without feeling deeply guilty. (For more discussion on the readability of essay manga, see chapter "Essay Manga".)

To make drawings easy to interpret, Hosokawa employed the method of visual exaggeration to some extent. "For example, I might have added a lot more shadow around the depressed Tsure. I probably tried to make the contrast between happy and unhappy times much sharper than they actually were in real life," she admitted. Perhaps that explains why the character Ten-san is sometimes drawn with an enlarged face with sweat beads or something comical that can easily catch the reader's attention. As a manga artist of some twenty years, Hosokawa knows the ins and outs of this business. "With no exaggeration, including the exaggeration of emotions, manga doesn't work," she clarified at first. "But the truth of the matter is, a person with severe depression really has huge mood swings. In that regard, manga is a perfect medium for telling a story about depression. By contrast, it's much harder to write a story about ordinary people with uneventful lives," she added.

Besides the issue of readability, she kept a few other points in mind while creating this series. For example, she made sure the people who appeared as characters were portrayed "appropriately so as not to hurt their feelings" even though, she admitted, it might not be possible to do so with every single individual because of the very nature of the subject matter as a mental health crisis. Another consideration is the importance of anonymity. She carefully tailored the way each character was drawn by avoiding the use of their exact verbatim, real-life identity, and physical appearance, so that they would remain unidentified, even by themselves if they read the manga. "The reason is I don't like anyone to end up feeling terribly embarrassed if they find their exact words in the story, for example," she added. This consideration was mentioned by another *mangaka*, Shiramizu Sadako, whose interview is featured in chapter "Shiramizu Sadako's *Uchi no OCD* (2015): A Collaborative Memoir of Obsessive-Compulsive Disorder".

Reception and Generalizability

As mentioned earlier, the first volume of *Tsure utsu* was an immediate blockbuster. The publisher, Gentosha, sold close to a million copies of *Tsure utsu*.[4] Because of the overwhelming reader response, the series' first volume was adapted into a television drama series and then into a feature film. "At first, I was confused because I had not expected it to be a best-seller, let alone drama and film adaptations," said Hosokawa in the interview. "It was tough emotionally." Her first surprise was the manga's high ranking since all she had hoped for was its "decent enough" media reception to secure more work for the future ("*tsigi no shigoto ga kuru kurai niwa urete hoshii*"), by which she meant that her goal was simply to meet the sales quota for reprints a few times ("*juhan ga nankaika kakaru gurai*"). That is why she was utterly taken aback when she discovered the volume's immensely positive reception.

To account for her Cinderella debut, Hosokawa pointed out three social factors that created a fertile ground for her manga's popular reception. First, NHK and other television channels had broadcast various programs about depression as a public concern. Thus, people were somewhat familiar with the term "depression" by then and would find a manga about a hardworking salaryman's mental meltdown to be believable. Second, the genre of essay manga had recently experienced a boom. To my question about the genre's sudden popularity, she responded, "The boom was ignited by Oguri Saori's manga, *My Darling is a Foreigner*. It sold in huge volumes. It was a mega hit." Oguri's essay manga was published in 2002 by Media Factory, one of the subdivisions of KADOKAWA publishing company, and had sold more than three million copies (*Pixiv Essay*, 2016). (As with many blockbuster manga titles, Oguri's manga was adapted into a film in 2010.) That was when Gentosha considered publishing manga titles in that genre. "The timing was perfect!" Hosokawa said. However, what made Hosokawa's manga popular within a week was the third factor: *Tsure ustu* was a unique manga in and of itself in two aspects, on which she elaborated further below.

First and foremost, *Tsure utsu* was the first graphic memoir of depression written by a family tōjisha. She said, "Except for some manga written by depression tōjisha,[5] no manga written by a family member existed prior to this work. That's what made it attract a lot of attention from readers, I heard." According to her, even her publisher had not realized how exceptional the manga was and was shocked to see it morphing into a smash hit. Back then, many family members were feeling at a loss, trying to figure out how best to interact with the depressed. That need created a niche for this manga to fill. "I've received many replies from readers telling me, 'This manga helped me find what attitudes I should have toward them!'" she said proudly. Other common reader feedback that she received was that the book allowed the tōjisha to see their condition more objectively and helped relieve them from guilt.

The manga was also unique for its boldness in having the word "depression" in the title. It was considered "taboo-breaking," according to Hosokawa. "Back then, publishers were apprehensive about using the word *utsu* (depression). But I used the very word in the title, which was fine with my *tantō-san* (my manga editor) as well as my publisher. Someone even complimented me, saying, 'That's a good title.' People at the other publishers told me, 'How gutsy you are!'" she added. Then, she went on to say, "At that time, people even avoided uttering the word *depression* as if it were an omen. The social atmosphere at that time was like, if you ever revealed to the public that you or your family have depression, you would be ostracized by society. That was the consensus back then." Again, this is reminiscent of how cancer patients were perceived by the public in developed countries even fifty years ago. Laughingly, she admitted that it was outrageous to include the name of the illness in the title because of that social climate. "Surprisingly, after my book was released, so many people came out saying, 'Well, to tell you the truth, me too.'" She added, "It [my manga] enabled people to openly state, 'I have depression' or 'My family has depression.'" In this regard, *Tsure utsu* broke the mold about the subject of depression. Readers spread the word about this amazing manga, and its sales skyrocketed.

It is important to stress that the importance of this manga's contribution to the public's understanding of depression cannot be measured by numbers only. As pointed out by Kodaira and Ito (2008), Hosokawa's manga enables those with no knowledge of the illness to relate easily with the characters Tsure and Ten-san; it informs the reader of how depression might affect family relationships and what family tōjisha might be able to do in dealing with the loved one's depression, an aspect that had been absent in the previous graphic memoirs on disability and illness. In addition to the value of the family's perspective, *Tsure utsu* offered a bona fide story about lived experience of depression, for which the public was hungry. As mentioned in the section of "Target Readership," Hosokawa was motivated to write this manga because Tsure's condition did not progress in the way many books on depression explained. Tsure read many books on depression from the library while he was still sick and shared as much information as he could with her all the time. ("Practically, that's how I gained all of my information about this illness," she said.) However, she came to realize that the information offered in many of these books did not match Tsure's condition. Thus, Hosokawa decided to pay close attention to what he was experiencing, rather than just following the books' advice for the patient. She added, "Yes, the fundamentals [of this illness] are the same, but there are quite a few individual differences regarding its symptoms and other things. I felt it necessary to teach people how diverse the experiences are."

However, she also pointed out in the interview that his case's uniqueness is not the strength of the manga; rather, its value is the rarity of insights from both the family tōjisha and depression tōjisha. In fact, readers found Tsure's depression-induced behaviors relatable, sending Hosokawa comments like "Yes, my family member is like that" after the manga was published. In this regard, she believes this manga, especially the first volume, is a good read on depression, “an illness that is hard to fathom especially for those who have no direct experience,” in her words. The manga also tells the reader how the mental health crisis changes the person and what sorts of burdens the family may have to shoulder under the circumstances. Toward the end of the interview, she said, “I highly recommend this manga to anyone who wants to understand what this illness is like.”

Notes

1. Interestingly, in his memoir published 2010, *Konna tsure de gomen nasai*, her husband wrote that although his wife kept telling him to rest and get well first, he had no idea how aggressively she sought more work in the manga publishing market to support themselves financially (Mochizuki 2010).
2. *Garo* is an avant-garde manga magazine that was popular in the 1970s but fell out of fashion by the mid-1980s.
3. Literally meaning “manga notes,” in which the word *pu* (notes) implies that they are visual signs akin to musical note, *manpu* are commonly used signifiers of manga semiotics, drawn as visual signs, not as words. They function as metaphors and represent a character’s emotions, psychological states, and mental conditions. *Manpu* signs are metaphorical expressions of anger, sadness, happiness, and so on.
4. According to a media impact data website, 950, 000 copies of the volume have been sold so far. Yamashita (2009) reports that the manga’s immediate sales were more than 250,000 copies.
5. Although Hosokawa did not provide the names of manga, *tōjisha manga* on depression prior to her *Tsure utsu* are Fujiomi Shuko’s *Seishinka ni ikō* (*Let’s visit a mental clinic!*) published by Bungei Shinjūsha in 2002 and *Minna genki ni yandeiru* (*Everyone’s ill lively*) by Kobunsha in 2004 as well as Azuma Hideo’s *Utsu utsu Hide nikki* (*Depressed Hideo’s Diary*) published by Kadokawa Shoten in July 2006, just a few months before Hosokawa’s manga. Azuma’s story covers his hardest hit period of depression from July 2004 through February 2005 in a gag manga style. Azuma also published *Shissō Nikki* (*Disappearance Diary*) in 2005, depicting his lived experience of suicide attempts, disappearance, and alcoholism, his major work that won three major awards in Japan immediately after it was published: the Japan Association of Manga Artists Grand Award (Nihon Mangaka Kyokaisho Taishō), the Agency for Cultural Affairs Media Art Festival Manga Division Grand Award (Bunkacho Media Geijutsusai Manga Bumon Taishō), and the Tezuka Osamu Culture Award Manga Grand Award (Tezuka Osamu Bunkasho Manga Taishō).

References

DSM-5 (*Diagnostic and Statistical Manual of Mental Disorders, 5th edition*). 2013. Washington, DC: The American Psychiatric Association.

Gendai Business. 2018. https://gendai.ismedia.jp/articles/-/56893?page=3 (Retrieved July 20, 2020).

Fujii, Katsunori. 2014/2016. *Nothing About Us Without Us: The Trajectory and Nature of Disability Rights Treaty* (*Watashitachi wo nukini watashitachi no koto wo kimenai de: Shōgaisha kenrijoyaku no kiseki to honshitsu*). Tokyo: Yadokari Shuppan.

Mochizuki, Akira. 2010. *Konna tsure de gomen nasai* (*Forgive me for being such a significant other*). Tokyo: Bungei Shunju.

Hata, Mikako. 2013. "Drawing 'Depression'—Examination of the potentials of comic representation." *Joshigaku Kenkyu/Konan Women's College Women's Studies Research Club 2*: March, pp. 66–75.

Hosokawa, Tenten. 2006. *My Significant Other Has Got Depression* (*Tsure ga utsu ni narimashite*). Tokyo: Gentosha.

———. 2009. *My Significant Other Has Got Depression Afterwards* (*Sonogo no tsure ga utsu ni narimashite*). Tokyo: Gentosha.

———. 2013. *My Significant Other Has Got Depression After Seven Years* (*Nananenme no tsure ga utsu ni narimashite*). Tokyo: Gentosha.

———. 2015. *Iguana's Bride* (*Iguana no yome*). Tokyo: Gentosha.

Josei no Hiroba, 2017. "Mental Autobiographical Comics Teach Us." August, pp. 102–113.

Kitanaka, Junko. 2012. *Depression in Japan: Psychiatric Cures for a Society in Distress*. New Jersey: Princeton Press.

Kodaira, Tomoe; Ito, Takehiko. 2008. "Mental Disability Memoirs—Multiple Meanings of Narratives [*Seishin shōgai no tōbyōki—tayōna monogatari no igi*]." *Macro Counseling Kenkyū* No.7: 48–63.

Kōno, Fumiyo. 2018. *Gigatown: Manpu zufu* (*Gigatown: Manga Signs*). Asahi Shinbun Shuppan: Tokyo.

Pixiv Essay. 2016. "Why is Essay Manga Popular in the Social Media? New Business Tactics of Publishers. [Essei manga ga SNS de ninki no riyu towa? Shuppangyokai mo aratana ugoki.]" 2016. June 13. https://ddnavi.com/news/305909/a/.

Yamashita, Keiko (2009). "Reading Comic Essay [*Komikku essay wo yomu*]." *Japanese Journal of Clinical Psychology*, 9(5): 698–701.

CHAPTER 9

Shiramizu Sadako's *Uchi no OCD* (2015): A Collaborative Memoir of Obsessive-Compulsive Disorder

Introduction

Obsessive-compulsive disorder, or OCD, is a mental disability manifested with obsessions (recurring and persistent thoughts and urges) as well as compulsions (repetitive and often very time-consuming behaviors such as excessive handwashing and ordering).[1] Individuals with OCD are not rare; approximately 2% of the general population is afflicted with this neuropsychiatric condition (Nakajima et al. 2018; Kobayashi et al. 2019). The condition is not simply an extreme form of cleanliness; it is a pathology because it causes significant distress and impairment in the person.[2] All compulsions that the person engages in are aimed at reducing anxiety, and yet its effect is short-lived, thus creating a vicious cycle: experiencing an irresistible urge, performing a meaningless act, and being released from the anxiety until the next urge arises. OCD is not only an illness that debilitates the person's functions but is also a mental disability characterized by high family involvement. OCD is one of twelve major mental health issues that affects thousands of citizens, just like depression and dementia, according to the website of the Ministry of Health, Labour and Welfare in Japan.[3]

Shiramizu Sadako's 2015 manga, *Uchi no OCD*, tells the reader about such an uncanny aspect of family life affected by OCD, an illness thousands of people still hide in public in Japan (Fig. 9.1). The author, Shiramizu, and her husband, whom I will call by his character name,

Y. Okuyama, *Tōjisha Manga*,
https://doi.org/10.1007/978-3-031-00840-5_9

Fig. 9.1 Image by Shiramizu Sadako. (Courtesy of Seiwa Shoten)

Ashika, collaborated on this manga as it is based on Ashika's OCD experience. In the interview, Shiramizu told me that her motivation for writing this manga was to equip herself with an instrument to explain "Ashika's mysterious behaviors," driven by his uncontrollable obsessions. By "instrument" (*dōgu*), she meant a method to provide legitimate explanations to the people around them (*mawari ni dōdō to setsumei wo tsukeru*). Without such an instrument to explain her husband's illness, she was frustrated and enraged by lay advice such as "It's just stress. Just take a vacation trip," "You'd better live in the quiet countryside," or "My relative got better with this method. You should try that," all of which can be offensive or outright harmful.

To draw the episodes of *Uchi no OCD*, Shiramizu had to elicit Ashika's direct feedback on certain issues through a method of research inquiry called *kikitori*. Some of his input was what he had never told anyone before. "Through this project, he taught me a lot about why he acted the way he did," she revealed. "I got to know him better by writing this manga. For instance, he told me his memory of his final moment with his father [a psychologically significant event for him] for the first time. I said to him, 'That's a great story. Let me use it for the manga! [*Iihanashijan. Itadaki dayo*!],'" she said with a little mischievous smile. Her reaction of "*itadaki dayo*" reminded me of the creative energy the other manga artists I interviewed had also experienced during the time of their personal crisis. These artists seem to possess the grit to turn a tragedy into an artistic triumph.

As stated earlier, *Uchi no OCD* is Shiramizu's collaborative work with her husband. While she is the main author, Ashika assisted her with both the content and drawings. In the interview, she elaborated on the processes of collaboration. The first step was her *kikitori* (face-to-face interview survey), a common method of inquiry in folklore studies in which the researcher obtains the informer's direct input through listening and note-taking. Clearly, without Ashika's cooperation with her *kikitori* task, this manga would not have represented the tōjisha's perspective. *Kikitori* is a fieldwork technique that she learned in her previous project, a children's book, for which she had interviewed local people and gathered folktales of *yōkai* (goblins). "In this approach, the investigator collects good material from others and puts them together," Shiramizu explained. She applied the same technique to gather Ashika's insights. This very method, according to her, is what separates this manga from regular, story manga. "I could embed my thoughts easily from a family perspective. But many things I had not known about this illness were revealed during this stage."

The second step was Shiramizu's preparation of the *nēmu* (draft manga with graphics and text, tentatively drawn in pen) using the notes of the *kikitori* from Ashika. Once the plot was developed, drawing begins, a task she called "*sakuga*." Ashika assisted her *sakuga* by drawing images such as landscapes, buildings, vehicles, and animals that she was not very good at herself. She would pass the draft paper to him, telling him what image she needed and where; once he finished drawing the requested image, he would pass the paper back to her so that she could finish it off with the characters and other drawings. By contrast, all of the human characters were drawn by her. Ashika elaborated, "It was a division of labor. She is good at drawing human figures, particularly detailed posture, so she drew those. Regarding my character, I told her in words to draw in this manner or actually posed for her [so that she could sketch]. I took care of the images I'm good at making [such as drawing cars, non-human figures, and landscapes]." Ashika continued, "So my participation was patchwork. From time to time, I'd respond to her questions [for the *kikitori*] or draw something like a car or a train station, whatever she had asked me to add." However, Ashika did not look at the entire manuscript of *Uchi no OCD*, because it was still painful to face his own illness at that time.

It was during these two stages that the majority of their collaboration took place. The third step was to bring the *nēmu* to her *tantō henshūsha* (manga editor) to discuss revision. The last step was completing the final version of the manuscript based on the editor's advice.

I asked Shiramizu if she had formally requested Ashika's permission to use his experiences. Her response was an informal consent from her husband. More specifically, she replied, "He wouldn't have told me in the first place, had he not wanted the story used in the manga. He knew my inquiry was for the book [manga]." I then asked Ashika if he had ever felt hesitant about her making manga out of his lived experience of OCD (as Hosokawa's husband did). He said, "It would be a lie to say I had no hesitation at all. But I wanted to confront myself and admit that I had an illness. With my condition, I can't hold a regular job, you know. But my drawing skills are useful [which helped this manga]. And I thought it'd be great if my experience could contribute to society in any way." Then, he added that he contributed because he was aware of the significance of this work. He also pointed out that more *tōjisha manga* have been published in the past ten years in Japan. Thus, it is "nothing unusual," in his words, to disclose one's mental health crisis in manga now. Still, putting oneself out there for all to see does not come easily, nor can it be done carelessly. Furthermore, it was unavoidable to disclose certain private matters of Ashika's childhood and family affairs to contextualize some of his obsessions and compulsions. They went ahead with such personal information, but quite importantly, they made sure the other characters in *Uchi no OCD* are unidentified in any way.

By disclosing his authentic story and showing what it is like to live with OCD via the popular medium of manga, Ashika himself wanted to help fellow tōjisha as well as their families. With mutual agreement, Shiramizu took her manga proposal and the first few sample pages to a couple of publishers. While the first company did not accept her proposal, the second one, Seiwa Shoten, responded with a resounding "Yes." *Uchi no OCD* was published as a takōbon book, unlike regular manga titles that typically begin as serialized in a print or digital magazine, prior to the book format. No prerequisite of magazine serialization is a unique style of Seiwa Shoten. "This publisher is also unconventional when it comes to scheduling. Maybe because they have so many book projects on their table, they don't demand you work with a strict deadline. They keep waiting until you're finished," she described.

Overall, it took Shiramizu approximately three years to complete *Uchi no OCD*. After she had created almost one-third of the work, the project went on hold due to the aftermath of the 2011 natural disaster which aggravated Ashika's OCD. "It was such a chaotic time in Japan. The images of the news covering the damage [by the earthquakes and tsunami in the northeastern area of Japan] were really shocking. We, too, felt the earthquakes [in Tokyo]. Many people were worried whether the water was safe to drink after the nuclear plant accidents. With all the commotion, we evacuated to my hometown in Izumo," she said. Excessively worried about possible concentrations of radioactivity in rainwater and other material, Ashika became completely incapacitated. The disaster also impacted Shiramizu psychologically, so "I couldn't advance the manuscript. I just had to leave it unattended for a while." Fortunately, however, their relocation to her mother's home in Izumo led to Ashika's regular visitation at a clinic, as shown in the manga (pp. 97–98 of *Uchi no OCD*). Together with the publisher's laidback approach to completion time, the wide range of periods from Ashika's childhood to his present days covered in the story also contributed to the prolonged length. "I was creating episodes ranging over twenty or thirty years, which is a long timeframe to cover," she said. The manga finally came out in print in 2015.

The Title and Target Readership

In Shiramizu Sadako's *Uchi no OCD* (2015), the word *uchi no* can be translated simply as "our" but contains multiple meanings, according to her responses during the interview. First, it is her collaborative work with her husband. Without his input, it would be impossible for her to portray accurately and vividly what it is like to live with this condition. Second, the author stressed that Ashika's symptoms may not be applicable to all OCD patients. Thus, by using the word *uchi no* (our/family's) in the title, she wanted to make clear that it is their experience.

Third, and most importantly, the use of the plural pronoun also implies that OCD is an illness that impacts everyone in the household. Both the sick individual and the family are trapped within the vicious cycle of obsessions and compulsions, and the illness becomes "ours" rather than "his" or "hers." For this reason, the title is appropriate even if it is Shiramizu's husband who is the OCD *tōjisha*. Shiramizu said, "One unique aspect of this illness is family involvement; everyone acts as if we had all contracted the same illness. [Living with Ashika], I, too, acted strangely on many occasions." As an example, she told me that to help ease Ashika's anxiety, she copied his way of handling the delivered mail: opening the envelope as though she were peeling a banana. "I directly experienced its influence on the family. It felt like I, too, had OCD."

Another example of comingling was that Shiramizu skipped an important work meeting because of Ashika's fear of rain contamination. Right after the 2011 earthquake-tsunami disaster, Japan was inundated with the disinformation about the rain carrying radioactive material from the damaged nuclear reactors in Fukushima. As explained earlier, Ashika was extremely afraid of that "contaminated" rain and avoided getting wet with even one drop of the rain. He wanted her to do the same. That year, Shiramizu was working on a children's ghost storybook and had to attend a monthly meeting at Popurasha, a publishing company in Tokyo. Ashika vehemently begged her not to go to work. Her sudden absence negatively affected her work status, of course. "At first, I didn't tell my editor and coworkers about his illness. They thought I was an irresponsible person (*katte na yatsu*). Although I finally explained my situation to the editor, my behavior must have seemed odd to the rest. His illness had some grip on me," she added. Ashika chimed in, saying, "I am aware that it can entangle everyone in the immediate family. My fear is not just for myself. I feel compelled to protect her too. That's how I made her do those things when my illness was full blown." (Later, Shiramizu called his reasoning "a twisted sense of protection," since it was not rational.)

Uchi no OCD depicts many other episodes that show she perfectly understands OCD as a "family pathology (***kazoku wo makikomu byōki***)," in her words. More than 95% of OCD households accommodate the patient's symptoms (Stewart et al. 2008). As Shiramizu stated, OCD traps both the sick and the family. Whether one has or does not have OCD symptoms, members of the same household tend to engage in the same behaviors that the person is compelled to make. Sadly, research on OCD shows family accommodation actually increases the severity of the symptoms, decreases the effect of treatment, and diminishes the quality of life of everyone involved (e.g., Grover and Dutt 2011). Similar to cases in Western countries, such family pathology is common among Japanese OCD patients and their families (Kobayashi et al. 2017). Following Japanese patients with OCD for ten years, Dr. Nakajima and his colleagues found that the more accommodation the patient receives, the longer the illness is prolonged (Nakajima et al. 2018). Several recent studies indicate that family intervention to reduce their accommodating behaviors in the form of cognitive-behavioral therapy (CBT) resulted in patient remission (e.g., Kobayashi et al. 2019).

The primary target readership of *Uchi no OCD* are family tōjisha—individuals whose loved ones are afflicted with OCD. Shiramizu aimed at these readers because she feels the most troublesome aspect of this illness is that the person's psychological issues are concealed (***me ni mienai nayami***) from the eyes of other members in the family. "If the disability is physical, it's clear to the family that the person needs help. But because this is mental, it tends to remain untreated until it gets worse," she said. Furthermore, odd behaviors of the afflicted such as checking the door multiple times and washing hands incessantly are not logically understandable by healthy family members, so the situation easily leads to family conflicts and strained relationships.

Shiramizu's secondary target readers are the tōjisha, many of whom are hesitant or unable to explain to others how they are affected by their own OCD symptoms. "This inexplicability makes the illness mystifying and obscure to the public. So, the ailment remains hidden within the family boundaries," she pointed out. Ashika followed, "There is some inconsistency about what causes my anxiety. Like, 'I can handle this but not that.' It doesn't make logical sense to me. I know it's absurd; it's meaningless.

But I feel compelled to act that way. It's as if I were possessed by my own [obsessive] thoughts. They are beyond my control."

Shiramizu also made sure not to use unusual imagery that might provoke stigma. For the tōjisha readers, she made certain that they would not find any part of the content offensive and avoided as much negativity as possible. Ashika said, "The reader seems to appreciate positive information more." Shiramizu elaborates, "People with OCD especially seem to like the view of the glass 'half filled' rather than 'half empty.' I found out that they like to hear creative and constructive tips on how to deal with challenging situations." As an example of such constructive strategies, Ashika brings his own chopsticks in case he is unable to use the restaurant's utensils because of his fear of germs. Shiramizu added, "In reality, though, things don't always go as planned. If we experience some failure, we tend to focus on that bad incident. But we need to have positive thoughts like 'I'm much better than before' or 'I can do this now that I couldn't do before.' More progress will be made if we focus on those."

Similarly, because mental health is a sensitive topic, she wanted to be extra careful not to stereotype the tōjisha or to cause misunderstanding about the symptoms described in the story. For that reason, she did not purposefully try to entertain readers or make them laugh. "Commercially speaking, I could have added more elements of entertainment, such as very funny episodes or more crude content like family drama. But I didn't," Shiramizu told me firmly. As reasons for that choice, she said, "For one thing, many of those who suffer from this illness are very delicate people. If the manga focuses on how comical their condition is, I'm afraid they will feel it's making fun of them." She did not intend to publicize the story simply to entertain the reader. Shiramizu said, "Fortunately, Seiwa Shoten is unlike other publishers that pursue profitability with popular, entertaining material." This company specializes in mental health-related publications written by psychiatrists and other specialists in the field. "So, the philosophy of this publisher worked to my advantage!" she added.

Because this manga primarily targets general, family readers who are more likely to be unaccustomed to reading manga, she made sure that the drawing style is easy for them to understand, neither too abstract nor aesthetic. Readability was her priority. For this reason, she stuck to using as many conventional manga metaphors as possible. She said, "I did the best I could not to revert back to my original drawing style. To make it readable, I completely removed my own artistic style." There, Ashika elaborated on how Shiramizu used to draw: "Her previous drawings were eccentric, in the style of *Ge ge ge no Kitaro.* She used pointillism, you know, drawing with black dots." Prior to writing *Uchi no OCD* in this genre of essay manga, her approach to "pursuing manga's artistry" was rather extreme. In those years, she was obsessed with that goal and immersed herself in perfecting that style. "Instead of drawing in a more efficient manner, I painstakingly added dots, one by one. Come to think of it, I doubt whether that pursuit had any meaning," she confessed. This manga enabled her to stop being obsessed with such esoteric goals and freed her psychologically from trying to perfect her artistic techniques. She explained, "Working on *Uchi no OCD,* I had to concern myself with issues such as how to best express abstract matters visually or what to avoid so that I would not hurt the reader's feelings. This experience helped change my artistic direction, in a sense."

The Author's Background

Shiramizu was the youngest among the manga artists whom I interviewed for this book. She was born in 1975, and as of July 2021 (our last correspondence), it is her fourteenth year as a professional illustrator. Unlike some of the other artists' inventive pen names, hers was an ordinary name comprised of her grandmother's first name, Sadako, and Ashika's grandmother's maiden name, Shiramizu. She told me, "Once I used this pen name, more work trickled in. So, I feel like my grandmothers are protecting me."[4]

As mentioned earlier, prior to *Uchi no OCD*, Shiramizu was working on a book on *yōkai* (goblins) for the series of children's literature run by Popurasha, a publisher specializing in children's books. It was a text-only book, not a manga, that involved fieldwork to collect local folktales of supernatural phenomena supervised by the late author Matsutani Miyoko.[5] Shiramizu and other fieldwork investigators met with Matsutani once a month and edited each other's writings. Prior to that project, she had written some comics including a short manga story about her hometown Izumo, a city in Shimane Prefecture, famous for Shinto folklore. But Shiramizu had no prior experience in writing literature or children's books, while the other members of her team were all from the field of folklore studies and had conducted fieldwork and written stories. Matsutani kindly told her, "Don't worry. We'll teach you the basics of how to write a children's book." From the team, Shiramizu learned the technique of the *kikitori* method and used it to learn about Izumo's local traditions and festivals from her mother. This experience later benefited Shiramizu in developing *Uchi no OCD*, where she used the same inquiry technique to obtain Ashika's input. "The whole process is the same. The only difference is the format of output—one is a book and the other, a comic," she commented. Furthermore, she enjoys discovering and expanding upon a story that already exists through research and then appending her drawings to the story, rather than inventing a story from scratch. "That type of story development seems to fit me the best," she said.

Shiramizu is a family tōjisha. Although she does not have OCD, she acted in a way that resembled her husband as he often begged her to accommodate his extreme germ phobia. Family members who end up enabling the patient, whether willingly or reluctantly, experience the illness's symptoms vicariously. Because this manga is her collaborative project with her husband, I will provide his background below:

Ashika is an OCD tōjisha. He is a soft-spoken man with a thoughtful demeanor. "About ten years ago, I sensed something was wrong with him," Shiramizu told me. One day, they stumbled onto a television program featuring OCD and learned that his excessive handwashing is a symptom of this illness. Unfortunately, Ashika was not psychologically ready to begin treatment at a mental health clinic back then. A nice, kind-looking man in his late thirties or early forties, Ashika shared this sentiment: "I was relieved rather than shocked at that time. It's because an illness can be treated but a personality cannot. I thought to myself, 'At least, something can be done,'" he said. Shiramizu felt liberated because the official label of an illness would enable her to explain logically to others—and herself—why he acted the way he did. "Before, his behaviors were inexplicable. But now that it's an illness, we can find reasons for it along with treatment for the problem."

People with OCD tend to find it hard to articulate to others why they are so compelled to engage in certain behaviors. Ashika commented, "Many of us are not even aware that our symptoms are part of an illness. But speaking for myself, I'd also say that we don't want other people to find out about our illness; we fear that they will think we are crazy. We are extremely careful not to reveal that secret, trying our best to hide the symptoms." Later in the interview, he elaborated on why people with OCD want to hide the fact that they are sick. *Monk* (*Meitantei Monku* in Japanese) is a famous American television show about a talented detective with OCD that was aired in Japan (2004–2010). He said in a soft voice,

"I kind of found the show detestable back then. I did recognize similar symptoms in the main character. But I was in denial; I couldn't admit that I had the same condition. I couldn't face it, especially when my symptoms were uncontrollable." Then, he added, "Now, I can probably watch the show without an emotional reaction." Shiramizu chimed in, "Their desire to protect their secrets from others, whether those are good or bad things, is probably much stronger than the average person. It seems to be related to their difficulty in facing the truth about their condition. Someone told me that he bought this manga but could not take it out of its package for a year." She added that people with OCD also tend to think that no one would understand them anyway, so they choose not to open up to others. "That's how they fall into the loop of not-telling and not-being-understood."

Fortunately, Ashika learned to confront his obsessions and compulsions. "Some of my symptoms are eased while others are not and continue to restrict my life. But I can go out more frequently than before. Like today, if I were still really sick, I would not have come to this interview. Recently, I became able to do more and more things," he said. Shiramizu followed, "Nowadays, he can accompany me more often to attend work-related meetings. He joins me in my Shinto study group." The study group meets in the Yushima Tenmangu in Tokyo, headed by Professor Ono, the shrine's *gon-negi* (junior priest) and also professor of religious studies, who teaches the *kojiki* in an approachable way to laypeople there. The group first studies the *kojiki* (an ancient collocation of Shinto traditions) together for about two hours. Then, they proceed to have a *naorai* (Shinto-style luncheon). "Sometimes, he can't eat together with us, and we have to tell the other participants, 'We need to skip the luncheon and go home today.' Because we have informed them of his illness, they will just say, 'Okay. No worries!'" she said with a smile.

Speaking of the *kojiki*, Ashika sees the origin of his germ phobia in some Shinto tales, albeit imperceptibly. "I began to participate in the study group for no special reason. But I came to notice some overlaps between the Japanese psyche (embedded in the collection) and my condition. We put much emphasis on purification (*kegare wo harau*)," Ashika stated. Shiramizu clarified, "Take that episode of the peaches from the Underworld, for instance. Once you visit the Underworld, you've been contaminated and must clean yourself. So, after returning from that world of deaths, Prince Izanagi performs a ritual to purify himself. That scene overlaps his [Ashika's] act of cleansing." She also describes a Shinto custom that originates in the concept of purification: *tokowaka* is a belief that Shinto buildings must be regularly renovated with fresh material. She explained, "That's why their shrines are demolished and rebuilt every now and then. Shinto values new and clean things and equates those elements with rejuvenation. It kind of tells you that the Japanese are, by nature, fastidious about cleanliness. Their pursuit of cleanliness is almost like an obsession. That essence seems condensed in the *kojiki* tales." Ashika added, "Understanding this ancient perspective is helpful in coming to terms with my predisposition to cleanliness. It also helps me understand where I am coming from."

Shiramizu emphasized that the treatment needs to be multi-faceted, as OCD cannot simply be fixed with hospital visits alone, nor can the medication address it single-handedly. In the manga, Ashika's character is shown to be taking Luvox (or Fluvoxamine in English; *depuromēru* in Japanese), a type of selective serotonin reuptake inhibitor, or SSRI, along with Lendormin to treat his insomnia (*Uchi no OCD*, pp. 104–106). Shiramizu speculates that his improvement cannot be accounted for by the prescriptions only and that another contributive factor is his visitation to a clinic where he talks about his worries and anxiety with a doctor who specializes in this illness. Although his clinic visits were not as frequent as they should be at the beginning, as his condition improved, he became able to keep up with his visitation schedule more easily. Shiramizu also attributed his progress to a technique of self-calming: after running into a situation that triggers his OCD reaction in a big way, Ashika gives himself a timeout until he feels calmer and his anxiety decreases. He admitted that using this "cool-down" technique appeared to be very helpful to him. Shiramize added that going out regularly, whether to a mental health clinic or somewhere else, also appeared to have some therapeutic effect on him.

Another effective approach to dealing with OCD-driven anxieties that Ashika told me about is CBT:

> I received several CBT sessions and learned that although anxiety increases at first, it dissipates as time passes and eventually disappears. At the peak of anxiety, if my hand touched the door by accident and I tried to relieve the anxiety by washing my hands, then, ironically, I'd get hung up with the peaked anxiety. But if I didn't do that and instead just held off and waited until my nervousness went away, it actually lowered my anxiety level. That's one of the key techniques of CBT, which works effectively with my panic attacks.

The CBT sessions Ashika attended were group-based and conducted by his physician, Dr. Sasa Takeshi, who also wrote the epilogue of *Uchi no OCD*.[6]

According to Shiramizu, each session at Dr. Sasa's clinic is about 600 yen (approximately USD 6) and is more affordable than a one-on-one, private session available at a medical university, which is about 10,000 yen (USD 100) per hour, but the patient must attend it weekly. Because Ashika's health condition fluctuates, he sometimes skips the group session. Still, he has learned some valuable techniques from the therapy. The session is attended by the tōjisha and the assisting physician only, and family members are not allowed to accompany them; what is discussed stays there and is not shared with anyone outside the group. "That is basically to protect the patients' confidentiality. If the family's participation is necessary, one can switch to a private session or join one of these family support groups," she explained. Although Shiramizu is interested in a family support group, she is hesitant about disclosing her experience and listening to other people's struggles. "I'm not yet comfortable with speaking in front of others," she added.

Ashika's condition still fluctuates, depending on the seasons. During winter, his health tends to deteriorate; in the spring, it improves. However, she pointed out, he would not have achieved the current level of recovery with the drug alone. Ashika told me that he had heard about the Exposure and Response Prevention therapy, or ERP (*bakurō hannō bōgai hō*)—*bakuro ryōhō*, for short—and would like to try it himself. It combines exposure with response prevention and is a type of cognitive-behavioral approach to treat anxiety disorders such as OCD, PTSD, and social anxiety disorder. "I've heard it's an effective method for people with OCD to manage their anxiety. In that method, I have to force myself to do a certain thing that I'm fearful of doing. Of course, I don't like to do that, but it's part of the therapy," Ashika commented.

Shiramizu was originally planning to work on a sequel to *Uchi no OCD*. At the time of our interview, they had a few topics already in mind, including therapies and rehabilitation opportunities. However, in our recent email exchange (July 2021), she wrote that she has run into issues such as privacy concerns about describing Ashika's group therapy, so she has not been able to move the sequel forward. Fortunately, however, this mental health topic continues to be covered by other *tōjisha manga*. Although I could not include Takahashi Shiki's (2018) manga due to the limited scope of this book, she shared some invaluable insight into her lived experience of OCD in our interview. Another tōjisha, Miyazaki Asuka (2018), also published two memoirs of OCD, *Kyōhakusei shōgai desu!* (*I've Got Obsessive-Compulsive Disorder*!) and *Kyōhakusei shōgai nikki* (*A Diary of Obsessive-Compulsive Disorder*) in 2018 and 2019 respectively.

Unlike Shiramizu's cute drawings of Ashika's compulsions, Miyazaki's manga metaphorically represents her OCD as a menacing, anthropomorphic ox man, an artistic strategy worthy of analysis on its own merits.

Shiramizu's Drawing Styles and Storytelling Characteristics

I asked about Shiramizu's drawing techniques to express imperceptible matters such as the character's emotions and thoughts, especially in response to OCD-related issues. She emphasized that Ashika's collaboration made her artistic representation of those matters possible through her drawing. "Especially in depicting what was mentally troubling him [Ashika], we worked collaboratively. I couldn't see inside his mind, so he explained it to me," Shiramizu explained. The more abstract the experience was, the more help Shiramizu needed from him. "It was like I played a receptive role at that time and just waited for his input patiently," she added. Using creative visual metaphors, Ashika described in words what it was like to experience an OCD attack. It was all because he had a habit of describing things like illustrations.

Ashika told me that he is fond of speaking figuratively and tends to use visual metaphors especially when he needs to provide concrete examples. He said, "Ever since I was little, I've had a communication complex and am on the outlook for good strategies. I've found it effective to apply metaphors inventively. This strategy also helps me figure things out for myself. That's why I use that." Shiramizu added, "During this project, when he was relaxed and calm, he was good at describing things figuratively. When he was well, he could think clearly and provided a lot of excellent visual metaphors for me. But when he was really sick, he couldn't even talk straight." This incorporation of various mental images that Ashika shared with his wife in the *kikitori* process made *Uchi no OCD* distinctive. To demonstrate the unique visual representations of *Uchi no OCD*, I provide the following examples.

Character Drawings

In *Uchi no OCD*, Shiramizu's character is a child-like, short woman named Kobuta-san who has a pigtail and black-framed glasses but no nose (see Fig. 9.2). This self-portrait is adorable but is a comically deformed version of the author. I asked her why she did not draw herself as pretty as she actually is. Laughingly, she responded, "It's a jocular version of me. I used to love playing dolls, so I can picture myself as a doll. It's kind of like operating myself in a puppet show." To contrast with the shy character of Ashika, she presented her character as a tomboy. (This technique is reminiscent of Nonami's.) "It's based on our real characteristics, but I exaggerated my 'feisty' personality a bit to make the contrast obvious, and perhaps, funny." Indeed, while screaming and crying, Kobuta-san speaks roughly in Japanese saying, "Why wash your hands every time! You say, this no good, that no good. Knock it off!" (first panel of Fig. 9.2). She even utters in anger, "You idiot! Go to hell!" (middle, right panel). As another episode of the manga illustrates, Shiramizu was not a student who would speak up for more obedient classmates in the art school where they met. She explained that her self-assertive nature stemmed from her growing up in a big family. "I learned to insist on my point of view. Otherwise, you'd be ignored [in a big family]. That's how I became self-assertive," she said with a smile. Interestingly, she pointed out, it was the shy Ashika who called out to her first at that art school.

As the episode of Fig. 9.2 narrates, when Ashika's OCD was severe and untreated, they were trapped in their world and could not find a way out. She said, "If someone in the family is really sick, you can't break out of the domestic situation on your own. If it's just you and your husband, both of you will go around and around in a vicious cycle." She stressed that breaking the cycle requires the help of external resources. "Medical professionals, or even someone you trust, needs to be there to get you to claw your way out of it. Once you're out, then the person's condition will improve, and you both can communicate better," she went on. Reaching that stage finally gave them some room to explore creative solutions to deal with his OCD compulsions.

Facial Expressions of Having a Panic Attack

In this manga, Ashika's panic attack is represented through his character's eyes. As shown in Fig. 9.3 (see left-middle panel of p. 96 and middle panel of p. 97), Ashika's eyes get a spiral pattern, unlike a Western drawing of bulged eyes and sweat to express a state of panic. "That image of a whirlpool was my idea," Shiramizu pointed out. However, the image of twirling eyes is a commonly used manga signifier. Using this conventional sign, she visually implied Ashika's incapacitated state [*zenzen dame na jōtai*] under the influence of OCD. She also expressed Ashika's poor mental state with a cross in one eye and a spiral in the other as in the middle

Fig. 9.2 Image by Shiramizu Sadako. (Courtesy of Seiwa Shoten)

Fig. 9.3 Image by Shiramizu Sadako. (Courtesy of Seiwa Shoten)

right-hand panel on page 97 in Fig. 9.3. Another hyperbolical expression of Ashika's anxiety is the wrinkled frame of the character's mouth (e.g., the middle panel on p. 97). Shiramizu explained, "He can't see himself. So, anything that can only be observed from outside, such as his facial expressions, is my idea."

Another example is the top right panel of page 72 in Fig. 9.4, in which a large eye in a whirlpool next to Ashika's character visually symbolizes his marked anxiety due to the trademark "Windows." Shiramizu drew that image based on Ashika's metaphor. "He told me it was like someone's peeking through one of the windows." As an OCD-triggered obsession, Ashika became fearful of the computer operating system, Microsoft Windows. To him, the name "Windows" collectively represents all computer operating systems. Because of a scary mental image of "a peeking eye" stuck in his mind, for a while, he was not able to use any computer with an Internet connection, whether it was a PC or Mac. He did understand that it was illogical to think that someone could "watch" him

Fig. 9.4 Image by Shiramizu Sadako. (Courtesy of Seiwa Shoten)

through the Internet. But that is what having OCD is like. The full episode of "The Windows Incident" is found in the manga on pages 69–72.

Ashika said that OCD feels like a spell put on a person who cannot dismiss obsessions and move on. "I know they [my obsessions] are ridiculous. But trying to ignore them only leads to increased anxiety," he explained. Not being able to use the Internet at home was a professional setback to Shiramizu when she began working for publishers in around 2006 because she had to electronically send her work and exchange emails with editors and clients. But because of Ashika's fear and strong rejection, she was unable to have her computer connected to the Internet. The situation forced her to conduct her business with an old-fashioned fax machine. "It was hard to explain to others what was going on at home. My writing was typed with and printed from the computer, and yet I couldn't send it to them electronically. They kept asking, 'Why not hook it up to the internet?' or 'How soon will you do that?' I gave them a vague answer like, 'I'm not sure,'" she said. "Thank goodness, it's over now. But

I still can't believe that it actually happened," she whispered. To show contrast, the emotional entrapment of Kobuta-san is represented through her pointy "blank" eyes (eyes drawn with no pupil), indicating her intensified anger in the right middle panel and the left bottom panel of page 73 in Fig. 9.4.

Cute Monster "Fuan" and a Monkey Train

According to Shiramizu, Ashika already had that image in his mind. He explained as follows: "When my anxiety level goes through the roof, my brain stops absorbing information properly. For example, I know the lock is on because I saw myself lock the door, but somehow, my brain refuses to retain that memory. I can't help checking the door again and again and can't leave home for a while. I try to convince myself that it's locked, but a part of me thinks, 'Wait, it might be unlocked.'" Shiramizu told me that once the anxiety attack happens, his brain freezes, and whatever others say doesn't get processed. By contrast, as Fig. 9.5 illustrates, when his OCD is relatively controllable, his anxiety is personified as a chubby, baby-like monster with a label *fuan* (anxiety) written on it (see two middle and two bottom panels on p. 107).

In addition, a monkey train is one of the recurring images in Ashika's mind, as seen in the middle and bottom panels of page 107 in Fig. 9.5. In that image, the monkey conductor, wearing a cap with the logo "OCD," operates the train carrying Ashika's character in it. The monster "Anxiety" appears to block the train's progression by sitting in the middle of the train track, as seen in the central panel of page 107. Once his medical treatment is underway, however, the train gains speed and zips by, pushing the monster, which has shrunk down in size, off of the track, as in the left middle panel. These were drawn according to Ashika's mental images. When he was describing them to Shiramizu, he actually used the term "monkey train" (*osaru no densha*). Understanding this metaphor requires some Japanese cultural familiarity. The monkey train was a common theme park attraction for children at one time in Japan.[7] Because it was a monkey, not a human conductor that operated the train, its progression was unreliable; yet, the child passengers enjoyed such a wacky ride. That illustration visually implies the precarious nature of living with unruly OCD symptoms.

107

Fig. 9.5 Image by Shiramizu Sadako. (Courtesy of Seiwa Shoten)

A Battle on a Remote Island

A unique representation of Ashika's battle with OCD is the image of his character stranded on a remote island. As shown in the top, right panel of page 114 in Fig. 9.6, a winged monster named "*Kyōfu no daimaō*" (Horror Demon King) flies over and attacks him. However, Ashika's character discovers the monster's weakness: an inflatable air cap above his tail (first panel of p. 115). The moment he succeeds in removing the cover and letting out the air, the monster shrivels away (middle panels). The image of the inflatable monster balloon is Ashika's idea, which he equated with conquering his obsession. The remote island is Shiramizu's image to symbolize his isolation by depicting him as alone in his world without being thoroughly understood by others. "I wanted to create a game-like scene in which the hero battles with the monster alone," she added.

114

Fig. 9.6 Image by Shiramizu Sadako. (Courtesy of Seiwa Shoten)

Hurdles to Jump Over

The image of Ashika jumping hurdles is a metaphor that Shiramizu collaborated on with him for this manga. In Japanese, the expression "unable to jump over because the hurdle is too high" (*hādoru ga takakute koerarenai*) implies the inability to overcome something. As Fig. 9.7 shows (see bottom panel of p. 127), the manga uses a visual image Shiramizu created by expanding upon Ashika's verbal expression about the high hurdle. (A similar image also appears in the top left image of Fig. 9.5.)

"Sometimes, he could give me a perfect description and at other times, he struggled," Shiramizu explained. "Abstract things such as fears and worries that only existed in his head needed his initiative, followed by our discussion. Otherwise, I could not figure out where to start," she added. Ashika also stated that through this sort of collaboration, he could test whether his imagery would be transparent to readers. "If they didn't make sense to her, they won't to the readers," he said. Shiramizu added laughingly, "This manga made us talk together to the max! What helped us get through is our mutual goal of publishing this as a manga book. We might have not spent that much energy had it been just a conversation intended simply to resolve a domestic issue, not to say that such a resolution would not be important. But because it had a clear goal and deadline, our motivation [to discuss it thoroughly] remained high. We never said to each other, 'Oh, let's not bother for now. Let's talk about it later.' The incentive to publish this book got us to talk our hearts out."

As she described these specific techniques of visual representations, Shiramizu explained why she believes that manga is the perfect medium of choice to reach out to the tōjisha. "We could have written a book, but we thought it would be best to use manga because many people with OCD find it hard to read, following sentences line by line. By comparison, manga has both pictures and texts, so we thought it was a better medium for OCD tōjisha." Research on reading and OCD reports that because of their compulsion to latch onto information, people with OCD tend to reread a particular sentence multiple times, a behavior that makes it hard to complete reading a single book (e.g., Abramovitch et al. 2013). Ashika said, "When reading, I sometimes get hung up on a certain part of the text and read it over and over and can't move forward. Other times, I lose concentration in the middle and go back to where I was. For instance, if I can't quite picture the landscape in my mind, I go back to that description to get a better grasp of it. I repeat the process of rereading like an

127

Fig. 9.7 Image by Shiramizu Sadako. (Courtesy of Seiwa Shoten)

obsession, and eventually I get tired of reading." Shiramizu hinted, "Imagine a record with a scratched groove. The record plays the same tune over and over. It's like that." Ashika also said, "It's not just reading. When watching a movie, I get that obsession as if someone in my head were telling me, 'Be mindful; go back and watch it more carefully.'"

This issue of readability demonstrates why essay manga is the perfect genre for narratives on brain and mental health. Shiramizu pointed out that she intentionally avoided the style of story manga, even though she used to create manga in that style. As mentioned earlier, she targeted family tōjisha as her primary readers. "I also wanted them to use *Uchi no OCD* as a starter for their learning about this illness." Thus, to her, the simple storytelling style of essay manga is the output format that fits her audience, who are likely to lack profound experience of reading such a genre of manga. As she explained, the story manga style typically contains plenty of textual information often presented in elaborate ways.[8] She said, "Japanese comics have more panels per page than Western comics in general. On top of that, a story presented in the story manga style usually has a complicated panel layout. It's not easy to figure out how the panels are sequenced [unless you have much experience with the style]." She pointed out that some story manga titles have a layout so as to make the reader stop and contemplate a particular scene. In other words, its complexity is embedded for an artistic effect. "I selected the simplest design of manga for the readers who are family members and have little or no manga reading background. I wanted to present the story in a relatively simple storytelling style," Shiramizu explained. That is why each page is organized, more or less, with only three layers of panels in *Uchi no OCD*. "It's much easier to follow the plot that way."

Reception and Generalizability

Shiramizu prefers not to discuss how many copies of *Uchi no OCD* have been sold. "Not a huge number to boast of anyway," she wrote in her recent email. Yet, she shared with me some of the readers' comments that she has received. The majority of audience feedback was positive, she added. Some messages were brief, like "To tell you the truth, my family member has it" and "Actually, I've been acting like that." The sick themselves are incapable of elucidating to others how their illness is affecting them. She explained that the lack of clear explanation frustrates family members and coworkers. One person wrote to her, "A colleague of mine behaves similarly, and this manga has helped me understand him better." Another wrote, "Thank you for letting the world know about this." Overall, she continued, family members or coworkers who have someone with similar symptoms benefited most from *Uchi no OCD*.

She has also received comments from OCD tōjisha readers such as "I received courage from this manga," "It helped me decide to get married," or "I was inspired not to give up." Some of them wrote that specific information, such as medicine-related information, Ashika's *hikikomori* (acute social withdrawal lasting longer than six months) experience, and his inventive ideas about dealing with obsessions and compulsions, was very useful. Other comments made her realize that people with OCD are still hiding their condition out of social stigma. These comments began to fill her social media feeds right after the manga was published (except for one letter mailed to her). She has both a Twitter account and a website where readers can post comments. She told me that it was most rewarding to see how much positive impact *Uchi no OCD* has made on people's lives.

OCD-driven fear manifests itself in different forms; thus, one memoir of OCD does not speak for every tōjisha. For this reason, *Uchi no OCD* mentions at the get-go that the symptoms described here are only relevant to "our case (*akumade uchi no bawai*)"; it is stressed that the reader should know that "the symptoms vary from person to person" (p. viii). For example, not every OCD tōjisha engages in heavy handwashing. In the same introduction, both Ashika and Shiramizu's characters speak to the reader saying, "Still, we hope our story will be of some use to you" (*oyakuni tateba saiwai desu*). As mentioned earlier, the manga is titled *Our OCD*, partly because OCD symptoms and degrees of severity are so diverse that they did not want to claim that their story represents every case. Shiramizu said, "The experience of anxiety is the same, but what they to do to avoid the trigger of anxiety is different." Ashika followed, "Each individual's OCD symptoms are unique. One patient finds doing a certain thing impossible while another patient has no problem with it. I myself wonder why this particular thing terrorizes me but not others. Also, there are cases where their obsessions are similar but show up in different compulsive behaviors."

"Even though how OCD is expressed differs from person to person, it is the same in that they are all more strongly compelled by obsessions than unaffected people are," Ashika said. The diversity of the symptoms also makes it difficult to describe OCD to lay people. "Because we can't present a so-called 'classic case,' it's hard to teach the general public what

OCD is. If we use one elusive example of obsession, the person will assume, incorrectly, that they don't have OCD." Due to this symptomatic multiplicity, Ashika and the other participants of the group therapy session were advised to just listen to each other's problems, while strongly discouraged from suggesting what others should do. However, they were encouraged to make the most of others' strategies if some of them were applicable to them.

After publishing *Uchi no OCD*, both Shiramizu and Ashika learned that there are quite a number of OCD tōjisha and their families who are still afraid to disclose the diagnosis publicly due to strong social stigma associated with mental illness. Ashika pointed out that the family shame of OCD contributes to the untreated condition as well as the phenomenon of extreme social isolation called *hikikomori*. "The government has not helped create many opportunities for people with OCD to congregate. I once Google-searched for a group and found only one or so per prefecture. We don't have many outlets to get together and talk with peers." Shiramizu added, "The medical system provides peer counseling opportunities. But outside of the medical services, you rarely find anything similar." She argued that because OCD is a psychological condition with which the public is relatively unfamiliar, such opportunities have not been created for the tōjisha. Furthermore, many people with OCD prefer being isolated to keep their "shameful" behaviors secret. "There are far more people who're suffering from OCD than you can imagine. Many of them hide the truth about their illness. If they hide it, people hardly ever know that they are suffering. In the end, the sick are left untreated," Shiramizu said. "Unlike the old times, they now send an SOS via social media. That's how we know people who have not come out publicly do exist. But they hardly ever get out of their homes and meet fellow OCD patients willingly." They do not need to suffer in silence, and yet many of them end up doing exactly that.

After this manga was published, Ashika became able to talk about his OCD with others. He recently completed the group CBT program and currently "manages himself well" without regular visits to the clinic. He has learned to apply techniques to handle his anxiety attacks and has changed his thought patterns to manage obsessions. Working as an employee at his wife's small business, he provides illustration services and helps create animation films for corporate and government clients. On top of that, he assists his wife in creating *kojiki* animation tales with shadow pictures.

There are also activities that he is still unable to perform. For example, his OCD-related obsessions prevent him from driving (due to his fear of running over people or animals by accident). He has a dream of driving a motorbike with his wife in the sidecar. "When I was a child, I was a fan of the hero *Kikaider* who rides in a sidecar in a television show. It'd be great if that dream came true one day. But honestly, that may not be realistic," Ashika uttered. Shiramizu encouraged him, "Realistically, you would first get well enough to drive a regular car." I smiled at them, picturing the adorable image of their riding the sidecar together.

Until recently, OCD had been one of the least publicly discussed mental health problems in Japan, which accounts for why there are only three other manga artists, Miyazaki Aasuka (2019), Takahashi Shiki (2018), and Ryu Tamako (2016), who have published OCD-related memoirs. Such rarity is appalling, considering the fact that 1–3% of the general population is affected by this disability (Nakajima et al. 2018; Kobayashi et al. 2019). Unlike the other more publicized mental disabilities such as depression and *hattatsu shōgai*, OCD manga does not seem to attract as many readers. However, I can attest that the person who gained most from *Uchi no OCD* is Shiramizu herself. James Pennebaker (2014) calls using the power of writing to speed one's recovery from challenging experiences "expressive emotions therapy." Shiramizu articulated that writing this manga enabled her to learn the ins and outs of the illness affecting her husband. She expressed her insight as follows:

> Unexpectedly, writing about my painful past [with his OCD] gave me a chance to heal myself. It was excruciating to recall those events. But to disclose my pent-up feelings through this work was definitely a catharsis, an emotional purging, almost like a therapy. I felt relieved afterwards. I was able to sublimate my feelings. (*shōkasaseru koto ga dekita*)

As mentioned earlier, this work also helped Shiramizu reset her artistic direction. "My effort to keep *Uchi no OCD* a reader-friendly manga for ordinary readers allowed me to see the benefits of simplicity. I found beauty in it. It was like resetting my mind completely." For this manga artist to have found a new direction early in her career must be both a personal gain and satisfaction, beyond fame or a commercial outcome. Furthermore, not only Shiramizu but also Ashika benefited from the collaboration, as their social network has expanded greatly. They said that, without this project, they would not have encountered so many interesting people beyond their regular confinement.

Notes

1. This description appears in the APA's DSM-5 (2013, p. 237). Similarly, the WHO's ICD-11 clusters obsessive-compulsive and related disorders as a group whose key symptoms are obsessions, intrusive thoughts, and preoccupations (https://icd.who.int/browse11/l-m/en#/http%3a%2f%2fid.who.int%2ficd%2fentity%2f1321276661). This entry is listed under the category of mental, behavioral, or neurodevelopmental disorders.
2. A common Japanese myth is that "a person who loves to clean" (*kireizuki na hito*) falls into this condition. To a lesser degree, people hold the myth that bad parenting causes OCD. Either one is a problematic view, since it goes against the medical definition of this illness.
3. https://www.mhlw.go.jp/kokoro/know/disease_compel.html.
4. In 2020, Shiramizu began to use her real name—her maiden name—professionally "to turn over a new leaf." Thus, her most recent works are published with the family name Kusaka, which has a root in Shinto folklore.
5. Matsutani Miyoko was a well-known author of popular children's books (e.g., *Little Momo-chan* 1974; *Tatsunoko Taro* 1978) in Japan. Shiramizu met the author in a free public lecture on *yōkai*. Knowing Shiramizu's Izumo manga story, Matsutani invited her to join her team of *yōkai* storywriters. That children's book became one of Matsutani's last works before she passed away in 2015.
6. Shiramizu said that Dr. Sasa had been very supportive of them through this project. As soon as she finished up her final version of *Uchi no OCD*, she took it over to his office and asked him to write an epilogue. Because Ashika began the CBT therapy at his clinic only after the manga had been written, there is no episode of the sessions.
7. Known as "*osaru no densha*" (monkey train), this popular attraction began at Ueno Park in Tokyo in 1948 and continued to operate until 1974 after Japan enacted a law to protect animals, the Act on the Welfare and Management of Animals (Shirado and Aoi 2008).
8. Since I teach a manga course myself, I can attest how challenging the story manga style can be to readers with limited familiarity with manga. I told her some anecdotes in which my American students find it extraordinarily difficult to understand manga titles designed in that style (e.g., Shimura Takako's *Hōrōmusuko* 2010).

References

Abramovitch, Amitai, Jonathan S. Abramowitz, and Andrew Mittelman. 2013. "The neuropsychology of adult obsessive-compulsive disorder: A meta-analysis." *Clinical Psychology Review*, 33: 1163–117.

DSM-5 (*Diagnostic and Statistical Manual of Mental Disorders, 5th edition*). 2013. Washington, DC: The American Psychiatric Association.

Grover, Sandeep, Alakananda Dutt. 2011. "Perceived burden and quality of life of caregivers in obsessive-compulsive disorder." *Psychiatry Clinical Neuroscience*, 65(5): 416–422.

Kobayashi, Yuki, Hisato Matsunaga, Tomohiro Nakao, Yuka Kudo, Eisuke Sakakibara, Ayako Kanie, Noriko Nakayama, Issei Shinmei, and Masaru Horikoshi. 2017. "The Japanese version of the Family Accommodation Scale for Obsessive-Compulsive Disorder: Psychometric properties and clinical correlates." *Journal of Obsessive-Compulsive and Related Disorders*, 15: 27–33.

Kobayashi, Yuki, Ayako Kanie, Atsuo Nakagawa, Yoshitake Takabayashi, Issei Shinmei, Noriko Nakayama, Keiko Yamaguchi, Chiaki Nakayama, Naotsugu Hirabayashi, Masaru Miura, and Masaru Horikoshi. 2019. "An Evaluation of Family-Based Treatment for OCD in Japan: A pilot randomized controlled trial." *Frontiers in Psychiatry*, 10 (932): 1–10.

Matsutani Miyoko. 1974. *Little Momo-chan*. Tokyo: Kodansha.

———. 1978. *Tatsunoko Taro*. Tokyo: Kodansha.

Miyazaki, Aasuka. 2018. *Kyōhakusei shōgai desu!* (I've Got Obsessive-Compulsive Disorder!) Tokyo: Seiwa Shoten.

———. 2019. *Kyōhakusei shōgai chiryō nikki* (A Memoir of Getting Treated for Obsessive Compulsive Disorder). Tokyo: Seiwa Shoten.

Nakajima, Akihiro, Naomi Matsuura, Keiichiro Mukai, Kyosuke Yamanishi, Hisashi Yamada, Kensei Maebayashi, Kazuhisa Hayashida, and Hisato Matsunaga. 2018. "Ten-year follow-up study of Japanese patients with obsessive-compulsive disorder." *Psychiatry and Clinical Neurosciences*, 72: 502–512.

Pennebaker, James W. 2014. *Expressive Writing: Words That Heal*. Idyll Arbor: Enumclaw, Washington.

Ryu, Tamako. 2016. *Kikakugai na fūfu: kyōhakushō otto to moto utsubyō zuma no hinichijō na nichijō*. (An Unorthodox Couple: Extraordinary events in an Ordinary Marriage between a Husband Suffering from Obsessive Symptoms and a Wife with Depression.) Tokyo: Takarasha.

Shimura, Takako. 2010. *Hōrōmusuko* (*The Wandering Son*). Tokyo: KADOKAWA.

Shirado, Takeshi, and Natsuki Aoi. 2008. *Why Did Children Stop Going to Amusement Parks?* (*Naze, kodomotachi wa yūenchi ni ikanakunattanoka?*) Soeisha: Tokyo, Japan.

Stewart, Evelyn S., Caitlin Beresin, Stephen Haddad, Denise Egan Stack, Jeanne Fama, and Michael Jenike. 2008. "Predictors of Family Accommodation in Obsessive-Compulsive Disorder." *Annals of Clinical Psychiatry*, 20 (2): 65–70.

Takahashi, Shiki. 2018. Basedōbyō ga gen'in deshita: Omake ni kyōhakusei shōgai mo. (The culprit was Basedow's disease [Grave's disease/hyperthyroidism]: And Obsessive Compulsive Disorder, too!) Tokyo: Bunkasha.

CHAPTER 10

Afterword

One day, I had a call from New York about a record-breaking surge in manga sales in the United States during the pandemic. "What do you think is behind it?" the voice on the other end asked.[1] Actually, manga's popularity is skyrocketing both domestically and internationally. The Publishing Research Office, *Shuppan Kagaku Kenkyūjo*, reported that in Japan in 2019, both print and digital sales of manga grossed approximately 498 billion yen, the equivalent of USD 4.3 billion, a 12.8% increase from 2018 (https://hon.jp/news/1.0/0/28155). Similarly, according to NPD BookScan,[2] more than 76% of sales in the adult graphic novels category in 2021 were generated by highly popular manga in the North American comics industry where manga sold a total of 24.4 million units in that year alone. All twenty titles of the top-selling comics in this category were highly popular manga titles such as *My Hero Academia*, *Demon Slayer*, *Jujutsu Kaisen*, and *Chainsaw Man* in the first quarter of 2021. By January 2022, the top winner was *Demon Slayer's* one-shot sequel *Demon Slayer: Kimetsu no Yaiba—Stories of Water and Flame* released by VIZ Media.

Manga is an unmistakably recognizable and economically important part of Japanese culture. In Japan, children and adolescents as well as university students and working adults read manga. Compared to its stigmatized status in the 1950s, manga had become more acceptable by the end of the 1980s in Japan, and a recent survey conducted by educators in

Y. Okuyama, *Tōjisha Manga*,
https://doi.org/10.1007/978-3-031-00840-5_10

Japan reports that children in grades 4–6 are more likely to have a positive perception of manga's educational benefits (Ookubo et al. 2020). School teachers and university professors utilize manga as educational tools in the classroom, as I discussed in Part I of this book. In 2000, *Mombushō*, or the Ministry of Education, Culture, Sports, Science and Technology (MEXT), published its white paper in which manga was officially acknowledged not only as an important form of modern expression but also as a legitimate artifact of Japanese culture. Annually, the MEXT celebrates the Japan Media Arts Festival where awards in several categories, including those of animation and manga, are presented to the creators. *Nihon Mangaka Kyōkai*, or the Japan Cartoonists Association (JCA n.d.) , also gives out the *Monbukagaku Dajinshō* (the Minister's Award from the MEXT) along with other awards for manga, a custom that began in 1990. Even the new prime minister Kishida Fumio tweeted recently that he completed reading the twenty-three volumes of Japan's most popular manga series, *Demon Slayer*,[3] a comment that attracted instant media attention. In the tweet, Kishida expressed his strong commitment to "support Japan's so-called soft power industries, including manga, anime, and film" (*Yahoo! Japan News*, September 2021). Due to the Japanese government's efforts using *Cool Japan*,[4] manga has been globally recognized as an economically influential commodity of Japanese culture. In the academic world, an academic organization, *Nihon Manga Gakkai*, or the Japan Society for Studies in Cartoons and Comics (JSSCC), was established in 2001. To respond to the global popularity of manga, Japan opened a manga museum in Kyoto in 2006, and the nation's first faculty of manga was established at Kyoto Seika University in the same year.

Regardless of the country of origin, comics are a subject of careful academic examination nowadays (e.g., Brienza 2016; Chute 2017; Kunka 2018; El Refaie 2019; Maekawa and Okumura 2020). But why are English-language comic readers' interests particularly strong in manga in North America now? A confluence of factors appears to have contributed to the record-breaking sales of manga. One factor is a crossover success between the twin sisters, anime and manga, as the majority of anime works are based on manga titles. As a primary factor, the success of anime adaptations from original manga appears to drive the overall manga sales in North America. For example, since the release of the anime *Jujutsu Kaisen* in October 2020, the sales of its original manga have increased exponentially both in print and digital from initial 8.5 million copies in October 2020 to 45 million copies in April 2021 (*Anime News Network*,[5] May 30,

2021). Prior to the manga boom, the demand for anime had already been growing for years overseas, partly due to the increased accessibility of watching anime imported from Japan; furthermore, streaming services such as Netflix and Crunchyroll have obtained more anime series for their platforms in recent years (*Worldwide*, May 24, 2021). The fact that anime was made more easily accessible by streaming platforms in 2018 and 2019 is a direct cause of the increasing popularity of anime internationally (*BBC* 2019). Although not always the case, the premiere of a new anime adaptation typically leads to an increase in sales of the corresponding manga as in *Demon Slayer: Kimetsu no Yaiba*. The sales for this manga series sporadically increased after the first season of its anime adaptation ended.

Another contributing factor frequently cited is the effect of the COVID-19 pandemic during which people were unable to go to the movies and enjoy social outings as much and had to seek other sources of entertainment. As subscribers of streaming platform companies begin watching the plethora of anime titles available in their database, new fans are brought to anime, which subsequently increases the sales of manga from which anime works originated (e.g., *CBR*, June 12, 2021). According to NPD BookScan reports, print manga sales in the United States increased by 3.6 million units in the first quarter of 2021 compared to the sales of the same period in 2020 (*Worldwide*, May 24, 2021). It appears people had more time to read manga during the pandemic, and a surge in the binge-reading of a manga series occurs after the success of its anime adaptation exposes the title to comic consumers (*TV Fandom Lounge*, December 24, 2021). Thus, it is undeniable that pandemic-driven lockdowns have influenced manga sales.

It is speculated that other factors leading to the sharp increase in manga sales in North America are consumer dissatisfaction with US superhero comics as indicated by their declining sales (e.g., *Worldwide*, May 24, 2021) and the entertainment quality of imported manga (e.g., *Hot Movies News*, August 11, 2021). The translated and imported manga titles are not only the tried and tested products that have already survived the scrutiny of Japanese consumers but are also the ones that are carefully selected by the North American comics publishers for Western audiences through their market research. In other words, these are the titles the North American comics industry has cherry-picked and had translated carefully by experts.

In fact, this new boom of manga has surpassed the peak sales of the previous boom in which manga sales continued to rise until 2007 while large-scale booksellers like Borders began to set up an entire section dedicated to manga titles in the United States. After the end of 2007, the profits of publishing companies plummeted abruptly because the market was severely influenced by the US financial crisis of 2008–2009. Anime was already popular in the United States in the 1970s while manga did not take root in the American turf until the 1980s, partially due to a "higher barrier for adaptation for language and format" (Saito 2019). By 2018, manga had outsold US superhero comics (*ICv2*,[6] July 29, 2019). Manga sales increased 350% from 2002 to 2007 and have morphed into a serious business as well as an art form that influences the larger world (Brienza 2016). Manga now appears to be "cool" even for the college-educated, American reader, perhaps symbolizing the Japanese government's dream come true. My own manga course gets filled immediately after it is released for enrollment. Fast forward to pandemic-ruled 2020 when North American manga publishers such as Yen Press and VIZ Media entered into an unprecedented time of record-breaking sales in a variety of genres from *shōnen manga* to *shōjo manga* to LGBTQ+ manga. Manga titles are now regulars on *The New York Times* Best Seller list. Avid manga readers no longer have to rely on scanlations (pirated digital manga translated by fans and amateur translators).

However, the importance of manga does not end with big name, budget-generating titles such as *My Hero Academia*, *Dragon Slayer*, and *Jujutsu Kaisen*. Non-blockbuster manga titles, such as the ones I introduced and analyzed in this book, have social resonance and educational significance. Especially in connection with mental health wellness, *tōjisha manga* are worthy of more attention from both academia and mass media. For instance, although the intersection between mental disability and autobiographical comics has been examined in recent publications, including *Visual Metaphor and Embodiment in Graphic Illness Narratives* (2019) by British scholar Elizabeth El Refaie and *Why Comics?* (2017) by US scholar Hillary Chute, Japan's *tōjisha manga* has not only been unrecognized but also left unexamined. Decentralizing such Western centrism in comic studies is the impetus for me to research and publish this book.

Manga as a Change Agent and a Tool for Mental Health Education

Manga, or comics in general, is a unique medium that combines text and image for storytelling. As brought up by these interviewed artists in Part II, manga has a strong appeal to the general public precisely because of this uniqueness, and under some circumstances, people experiencing a mental health crisis may find this medium easier to read than a text-only book. Take Ashika's case in chapter "Shiramizu Sadako's *Uchi no OCD* (2015): A Collaborative Memoir of Obsessive-Compulsive Disorder", for instance. His symptoms, especially obsessive thoughts, caused difficulty in comprehending text-only materials when his condition was most severe. Similarly, the author of the *Gaki-tame* series from chapter "Okita Bakka's *Gaki-Tame* Series (2011–2013): A Memoir of a "Troublemaker" Aspie Girl" told me that readers with *hattatsu shōgai* like her are more likely to absorb information from manga than books. Because of the ubiquitous presence of manga in Japan, people find this medium more approachable in general and benefit both from learning about the reality of mental ill health while still enjoying the narrator's humor and artistic creativity. Furthermore, as the authors of the *Akira-san* series ("Nonami Tsuna's *Akira-san* Series (2011–2017): A Memoir of a "Cassandra" Wife") and the *Tsure utsu* series ("Hosokawa Tenten's Tsure utsu Series (2006–2013): A Couple's Lived Experience of Depression") pointed out, people who would not read a technical book about depression or Asperger's syndrome may casually pick up a manga and learn about the condition of their spouses, family members, or coworkers. Once the manga becomes the talk of the nation, the general public's awareness and knowledge of the condition expands even more, which eventually and inevitably influences the policies of the government, schools, and workforce. In that sense, manga can be an agent for social change.

In medical books and articles, patients' lives get buried under statistics and diagnosis-based anecdotes. By comparison, memoirs illuminate these "faceless" individuals and humanize their lived experience. As I have been arguing in this book, *tōjisha manga*—graphic memoirs of life writing—humanizes the otherwise invisible disability of a neurological or psychological condition precisely because the first-person narrative makes one's lived experience more relatable and authentic to the reader. The strength of *tōjisha manga* lies in this power of humanization. As delineated in chapter "Tōjisha", tōjisha perspectives are vital in helping improve lay people's mental health literacy. Chapters "Tōjisha Narratives"" and "Essay Manga" also demonstrate that their stories, including *tōjisha manga*, can be used as a great teaching tool for schools and programs that train the next generation of health care and welfare workers. Unlike decades ago, manga is clearly penetrating North American and European communities of graphic novel readers now. Thus, educators can utilize this medium as an instructional tool to discuss important issues beyond Japan. As the disability studies scholar Ellen Samuels advocates, graphic memoirs are not only accessible but also compelling teaching materials suited for American university undergraduates[7] especially in courses on mental disability (Samuels 2015). For these reasons, *tōjisha manga* are even more deserving of scholarly attention. While journalists and business market analysts crunch sales numbers and examine the economic impact of manga, we academics have a job to do: investigate and demonstrate that manga can be a change agent for important issues such as mental health.

Manga is also a mirror into societal attitudes toward an issue. As demonstrated in Part II, the manga author's real-life interactions with their friends, family members, and other lay Japanese, we vicariously experience the ramifications of having a stigmatized condition such as depression. Through their artistic works, manga authors tell us about real-life prejudice and discrimination against the affected individuals and their families as well as the common misperceptions and unfounded exaggerations toward a certain behavioral health problem at work and in the community. As a lifelong caregiver of my own mother with OCD, I personally thank these authors for disseminating their knowledge and dispensing advice and caution learned through their debilitating experiences. It is my belief that, if translated, *tōjisha manga* titles are an indispensable resource for English-speaking readers to understand how lay people react to mental and brain health issues and how each specific condition is treated by medical professionals and assisted by the government in contemporary Japan.

This book is also an ode to the courage and valor of these mangaka who dared to share true stories despite many risks associated with one's disclosure of a medical condition, particularly because most of these manga artists were unknown before their *tōjisha manga* titles garnered some celebrity status.[8] Furthermore, reliving the experience by writing about it in manga can also be excruciating, especially if the author's experience includes serious school or workplace bullying, shunning by friends and in-laws, and suicidal attempts and other self-harm. In this sense, the authors of tōjisha manga are not just survivors but also conquerors of a hard lot. They are worthy of our attention and admiration.

Mental health was always an important social topic but has become even more so since the world was stricken with the COVID-19 virus. The pandemic-driven isolation and acute economic downturn have impacted many Japanese, especially women, causing a spike in depression diagnoses; more people were reported to have died from suicide than the virus itself by the fall of 2020 (e.g., Shigemura and Kurosawa 2020; Ueda et al. 2020). In the United States, just when we felt we were about to get over the hump, Omicron, the new variant, has thrown a wrench into everyone's plans, and our lives continue to be disrupted and restricted. At the time of writing this, the number of cases appears to be declining. Who knows when another variant of COVID-19 will pop up and derail our lives yet again. Even if we are fortunate enough to get over this, the emotional, physical, and economic damages will linger for a while after our journey of recovery commences.

All things considered, the two interwoven themes of this book will remain socially resonant for a while as we gradually recover from the pandemic's psychological impact and as manga continues to function as an integral member of Japan's soft power.

Notes

1. The voice belongs to Jim Zarroli, a former correspondent of the National Public Radio (NPR) covering business and economics. Writing a piece about manga sales for NPR, he called me on February 10, 2022.
2. NPD is a business that provides business data and analysis in various industries including books, apparel, footwear, and fashion. NPD BookScan (https://www.npd.com/industry-expertise/books/) publishes sales data and market analysis for the books industry. According to their website, their data covers approximately 85% of the US trade print book market, ranking best sellers per piece sales.
3. Asahi Newspaper EduA (2020), an online news site that provides education-related news and interviews, identifies *Demon Slayer: Kimetsu no Yaiba*, as the manga title most contributive to the spike and states that the manga's popularity brought a recent boom for *kendo* (sword martial art) practice and *kanji* characters studying among young manga readers.

4. *Cool Japan* is a slogan representing the Japanese government's efforts to promote overseas the recognition and popularity of cultural commodities such as film, manga, anime, video games, fashion, music, and cuisine. The efforts were spearheaded by the Japanese Ministry of Foreign Affairs in 1980. In 2010, the Creative Industries Promotion Office was established as its "strategy" sector within the Ministry of Economy, Trade and Industry (METI). The chief minister responsible for the operation of the sector was selected in 2012, generating a series of strategic plans and initiatives from 2015 to 2018 with the collaboration of businesses in the private sector (Cabinet Office 2019). The governing body of the *Cool Japan* efforts involved not only METI but also several other ministries, including the Ministry of Education, Culture, Sports, Science and Technology (MEXT). Simultaneously, a public-private fund, Cool Japan Fund Inc., was also founded in 2013 and began to execute "soft power" efforts (Cabinet Office 2019). Interestingly, the Fund invested USD 30 million in a licensor company to increase the presence of anime in the US market in 2019 (Anime News Network 2019).
5. *Anime News Network* is an online anime news site that reports news about anime, manga, J-pop, and other Japanese pop culture icons found in North America, Australia, and Japan.
6. *ICv2* is an online trade magazine covering geek culture news and sales analyses about popular culture including anime, comics, and games.
7. I have been using manga in my course, *Gender and Disability in Manga*, at the University of Hawaii at Hilo for more than ten years. My American students who grew up reading manga and watching anime enjoy reading about disability through the perspectives of manga's protagonists. Many of them have confided in me that it was an eye-opening experience as they read beyond "entertainment" manga like *Hunter x Hunter*.
8. In comparison, celebrity admission of mental illness may be more likely to be taken positively because of the person's star status. When Naomi Osaka, one of the world-class, female tennis players, recently revealed her "long bouts of depression" (*The New Yorker*, June 1, 2021), many of her fans and colleagues responded with sympathy and support. Her brave disclosure appeared to have helped break the mold and lower the stigma of mental disability in the international athletic world. However, the stakes are probably much higher when such a private matter is divulged by less famous individuals, especially in a country like Japan where depression is incorrectly associated with the loss of the person's mental self-control (The Lancet 2002).

References

Asahi Newspaper EduA. 2020. Manga's influences on children. December 24, 2020. https://www.asahi.com/edua/article/14031290 (retrieved February 19, 2022).

Anime News Network. August 1, 2019. "Cool Japan Fund Invests in North American Anime Group Led by Sentai Holdings." https://www.animenewsnetwork.com/press-release/2019-08-01/cool-japan-fund-invests-in-north-american-anime-group-led-by-sentai-holdings/.149602 (retrieved February 10, 2022).

Anime News Network. May 30, 2021. Cited in "Why is Manga So Dominant in American Graphic Novel Sales?" https://www.animenewsnetwork.com/news/2021-05-30/jujutsu-kaisen-manga-circulation-jumps-590-percent-to-50-million-copies-since-anime-debut/.173399 (retrieved February 12, 2022).

BBC (British Broadcasting Company). November 11, 2019. "Is Japanese anime going mainstream?" https://www.bbc.com/future/article/20191127-the-spirited-world-of-japanese-anime (retrieved February 19, 2022).

Brienza, Casey. 2016. *Manga in America: Transnational Book Publishing and the Domestication of Japanese Comics*. New York: Bloomsbury Academic.

CBR. June 12, 2021. "Japanese Media Is More Popular Than Ever—So Why Are Manga & Anime Still 'Niche'?" https://www.cbr.com/japanese-media-anime-popular-niche/ (retrieved February 12, 2022).

Chute, Hillary. 2017. *Why Comics?: From Underground to Everywhere*. New York: HarperCollins.

El Refaie, Elisabeth. 2019. *Visual Metaphor and Embodiment in Graphic Illness Narratives*. New York: Oxford University Press.

Hot Movies News. August 11, 2021. "Why the US Manga Market is Booming." https://hotmoviesnews.com/2021/08/01/why-the-us-manga-market-is-booming/ (retrieved February 15, 2022).

ICv2. July 29, 2019. "Big Split in Graphic Novel Sales in 2018: Kids and Manga Up, Superheroes and Author Titles Down." https://icv2.com/articles/news/view/43718/big-split-graphic-novel-sales-2018 (retrieved February 19, 2022).

Japan Cartoonists Association (homepage). n.d. https://nihonmangakakyokai.or.jp/about/about07.

Kunka, Andrew J. 2018. *Autobiographical Comics*. New York: Bloomsbury Academic.

The Lancet. 2002. "Stigma of Mental Illness in Japan." Vol. 359: 1866.

Maekawa, Osamu, and Hiroshi Okumura. 2020. *Manga/Manga/Manga: Jinbungaku no Shiten kara* (*Manga/Manga/Manga: From the Field of Humanities*). Kobe University Press: Kobe, Japan.

The New Yorker. 2011. "Naomi Osaka's Complicated Withdrawal from the French Open." (written by Louisa Thomas) June 1, 2021. https://www.newyorker.com/sports/sporting-scene/naomi-osaka-takes-a-complicated-stand (retrieved December 1, 2021).

Ookubo, Noriichiro, Hirokazu Wada, Shunichi Kubo, and Ryuya Horida. 2020. "*Shōgaku kōgakunen jidō ni okeru manga dokusho* (*Manga Reading Attitudes by Grade 4–6 Students*)." *Japanese Journal of Educational Media Research* 27(1), 13–29.

TV Fandom Lounge, December 24, 2021. "Top 25 Best-Selling Manga of 2021; Jujutsu Kaisen, Tokyo Revengers & More." https://www.tvfandomlounge.com/2021/12/24/top-25-best-selling-manga-2021-jujutsu-kaisen-tokyo-revengers-more/ (retrieved February 12, 2022).

Saito, Satoru. 2019. "Why are manga outselling superhero comics?: Rutgers professor explores varying themes in Japanese and American comics." (An interview by Cynthia Medina at Rutgers University). Faculty Excellence. https://www.rutgers.edu/news/why-are-manga-outselling-superhero-comics (retrieved February 17, 2022).

Samuels, Ellen. 2015. "Picturing Mental Disability in the Classroom." *Disability Studies Quarterly*, vol. 35, no. 2. https://dsq-sds.org/article/view/4865/3949 (retrieved December 8, 2021.)

Shigemura, Jun, and Mie Kurosawa. 2020. "Mental Health Impact of the COVID-19 Pandemic Japan." *Psychological Trauma: Theory, Research, Practice, and Policy*, 12 (5): 478–479.

Yahoo! Japan News. "'Kimetsu no yaiba' *dokuryō no Kishida Fumio shinsosai*" (The new prime minister Kishida Fumio's completion of reading 'Demon Slayer'). September 2021. https://news.yahoo.co.jp/articles/e01a26d3a-c7ae29a058813e59ad8eff5a60031ff (retrieved February 14, 2022).

Ueda, Michiko, Rober Nordstrom, and Tetsuya Matsubayashi. 2020. "Suicide and mental health during the COVID-19 pandemic in Japan." *Journal of Public Health*: 1–8.

Worldwide. May 24, 2021. "Worldwide, manga is outselling American comics." https://ultramunch.com/worldwide-manga-is-outselling-american-comics/ (retrieved February 13, 2022).

Index[1]

A

Adult Asperger's syndrome (*otona no asuperugā*), 14, 145

Akira-san wa asuperugā (Husband, Akira, Has Asperger's Syndrome), 11, 14, 124, 145, 148

Ashika, 251–263, 265–267, 269, 270, 272–275, 276n6, 283

Asperger's syndrome (AS) (*asuperugā*), viii, 11, 14, 25, 26, 42, 43, 113, 115, 116, 121–125, 127, 136, 142n4, 147–154, 156, 157, 159, 162, 169, 170, 173n2, 174n3, 174n4, 174n6, 174n7, 175n14, 283

Attention deficit hyperactivity disorder (ADHD), 11, 113, 115, 123, 124, 136, 142n1, 175n14

a-un no kokyū (a-un breathing), 170

Autobiographical comics, 2–5, 8, 10, 11, 13, 22, 43, 51, 66–69, 71n10, 77–103, 142, 173, 282

Azuma, Hideo, 81, 87, 92–95, 102, 196, 199, 207, 213n4, 213n6, 228, 246n5
  Shissō nikki, 87, 92, 93, 95, 102, 196, 207, 213n4, 213n6, 228, 246n5

B

Baikai-sha (go-between/intermediary agent), 78, 79, 83

Bakka-chan, 106n15, 116, 117, 120, 122, 127, 129, 130, 133, 135, 137, 228

Bethel House (*Beteru no ie*), 26, 34–42, 44n3, 46n10, 48n17, 95, 193

C

[the] Cabinet Office (*Naimushō*), 1, 15n2, 286n4

"Cancer of the mind" (*kokoro no gan*), 203

[1] Note: Page numbers followed by 'n' refer to notes.

Y. Okuyama, *Tōjisha Manga*,
https://doi.org/10.1007/978-3-031-00840-5

Caregiver/family *tōjisha/kazoku tōjisha* ("family *tōjisha*"), 11, 26–28, 43, 56, 59, 106n13, 217, 225, 226, 228, 242, 244, 245, 255, 258, 272
Cassandra syndrome (*Kasandora shōkōgun*), 11, 14, 145, 147–151, 153, 154, 156, 158, 159, 168–172
Chibi-characters/*chibi*-style drawing, 196
Chibi Maruko-chan, 79–82
Chōkaku kabin (auditory hypersensitivity), 123, 129, 130
Cognitive behavioral therapy (CBT) (*ninchi ryōhō*), 34, 223, 255, 260, 274, 276n6
Communication breakdown/ communication problems/ communication difficulties/ communication complex, 136, 147, 153, 158, 170, 171, 262
Cool Japan, 280, 286n4
COVID-19/the pandemic, 1, 4, 15, 172, 203, 279, 281, 285

D

Deforume (deformation), 80, 129
Deikea sentā (daycare center for mental health patients), 65
Depas, 190
Depathologizing, viii, 55
Depression
  major depression, 11, 71n8, 173
  and personality, 190, 221, 222
  utsu (depression), 96, 183, 217, 219, 244
  utsubyō (depression illness), 11, 183, 217, 219, 234
Developmental disabilities (*hattatsu shōgai*), 11, 14, 34, 113, 115–120, 123–125, 129, 130, 133, 136, 137, 139, 140, 142, 142n1, 142n2, 142–143n4, 143n5, 143n7, 171, 175n14, 275, 283
Diagnostic and Statistical Manual of Mental Disorders, 5th Edition (DSM-5), 10, 12, 16n6, 55, 113, 142n1, 147, 153, 183, 222, 231, 276n1
Dirty-joke gags (*shimo neta gyagu*), 188
Distress
  mental distress, 2, 5, 55, 57, 58, 227
  mental health distress, 51
Dōsen (motion lines), 159, 165

E

Early intervention (*sōki shien*), 59, 61
Enigmatic four-syllable word (*nazo no yonmoji*), 183, 221
Essay manga/*essay manga*/comic essay, 1–3, 6, 13–15, 40, 69, 77–103, 121, 125, 149, 150, 155, 156, 159, 168, 170, 173, 174n9, 175n10, 202, 217, 221, 225, 228, 242, 244, 257, 272, 284
Exposure and Response Prevention therapy (ERP) (*bakurō hannō bōgai hō*), 261

F

Family pathology (*kazoku wo makikomu byōki*), 255
Feeling of "something is out of place" (*iwakan*), 165
First-person narratives, 2, 3, 13, 51–55, 84, 283

Fuan (anxiety), 267
Futōkō (truancy; refusing to go to school), 123, 140
Fūzoku (adult entertainment), 119, 136

G
Gag manga/gyagu *manga* (hyper-comedic comics with slapstick jokes and funny plots), 6, 80, 81, 85, 91, 92, 97–102, 116, 120, 129, 136, 142n2, 174n9, 185, 187, 188, 208, 213n6, 228
Gaki no tameiki, 11, 14, 113, 116, 136, 175n12
Gekiga-style, 188
Genres (of manga), 3, 4, 6, 13, 15, 56, 69, 77–80, 84–88, 91, 104n6, 104n8, 116, 120, 125, 150, 155, 168, 174n9, 175n10, 187, 188, 208, 213n6, 223, 244, 257, 272
Graphic memoirs of illnesses and disabilities/graphic pathographies, 67, 68

H
Hattatsu shōgai (developmental disabilities), 11, 14, 34, 113, 115–120, 123–125, 129, 130, 133, 135–137, 139, 140, 142, 142n1, 142n2, 142–143n4, 143n5, 143n7, 171, 175n14, 275, 283
"Healing-type" (*iyashi-kei*) characters/"wobbly characters" (*yurukyara*), 91, 159
Hikikomori (acute social withdrawal lasting longer than six months), 119, 143n6, 273, 274
Hi-tōjisha (non-*tōjisha*), 23, 35, 60, 62, 64, 65, 71n6, 103, 185
Hiyaase wo kaku (break into a cold sweat), 232, 238
Hosokawa, Tenten, 7, 8, 11, 14, 65, 79, 85, 92, 93, 106n13, 170, 173, 174n8, 184, 186, 203, 208, 212n2, 217, 222, 225, 242, 252, 283
Humor, 13, 40, 41, 63, 69n1, 78, 80, 84–86, 90, 92, 96, 97, 99–103, 105n10, 120, 122, 168, 206–207, 228, 241, 283
Hypersensitivity, 123, 129–133, 141

I
Ibasho (place of belonging), 65
Ikizurasa (difficulty in living), 47n16, 85, 139, 143n4, 153, 170
International Classification of Diseases, 11th Edition (ICD-11), 10, 12, 16n6, 59, 142n1, 147, 276n1
Itako mangaka, 189

K
Kakimoji or tegaki moji (handwritten words), 186
Kaneko-kun, 199, 202
Kankai (remission) *vs. kanchi* (cure), 192
Kanshaku-suji (anger marker), 232
Kikitori (face-to-face interview survey), 251, 252, 258, 262
Kobuta-san, 262, 263, 267
Kodansha, 6, 113, 155, 183, 189
[the] kojiki, 259, 260, 274
Kokoro no byōki (a sickness of the mind), 64
Komame, Daruma, 96–102, 225

Kotatsu (Japanese electric foot warmer), 199
"Kyōfu no daimaō" (Horror Demon King), 269
Kyōhakusei shōgai (OCD), 11

L

Ladies comics (mature women's comics), 6, 155, 174n9
Learning disability (LD), 113, 115, 119, 125, 137, 175n14
Life writing, 13, 32, 35, 52–57, 69, 71n11, 283
Luvox (or Fluvoxamine), 231, 260

M

Madness/*madness*, vii, 9, 30, 31, 33, 38, 44n5, 55, 56
Mad studies/*mad studies*, 3, 22, 30–33, 38, 39, 44n4, 45n5, 45n7, 45n8, 46n13, 55
Mangaka (manga artist), vii, 2, 4, 6, 10, 15n4, 29, 56, 60, 61, 63, 78, 79, 81–83, 85–89, 91, 92, 96–100, 105n10, 105n11, 106n15, 113, 118–121, 125, 139, 142, 148, 154–157, 159, 162, 165, 168, 173, 185–191, 206, 208, 209, 213n5, 213n6, 217, 219–225, 228, 241–243, 251, 257, 275, 284
Manga kigō (signs used in manga semiotics), 129, 228, 246n3
(Manga) semiotics, 8, 104n6, 129, 189, 228, 232, 237, 242, 246n3
Manga's reading directions, 6, 8
Manpu, 8, 232–234, 246n3
Marital relationships, 158
Melancholy, 237
Mental disability/psychological disability, 1–5, 9–13, 15, 15–16n4, 21, 22, 24–26, 29, 30, 32, 33, 36, 38, 40, 48n18, 51, 54, 55, 59, 64, 70n3, 88, 92, 93, 95, 100, 142n1, 143n7, 191, 249, 275, 282, 284, 286n8
Mental health consumers/psychiatric system survivors, 9, 13, 21, 30, 55
Mental illness, vii, 9, 21, 30, 38, 42, 45n8, 46n11, 48n17, 55, 71n6, 115, 147, 156, 183, 186, 196, 202, 207, 208, 227, 274, 286n8
"The mind cold (*kokoro no kaze*)", 241
Miyao Masutomo, 149, 156, 158, 168, 171, 174n7
Mombushō, or the Ministry of Education, Culture, Sports, Science and Technology (MEXT), 280, 286n4
Monbukagaku Dajinshō (the Minister's Award from the MEXT), 280
Mondai-ji (problem child), 14, 113, 116, 118
Monk, 258
Monkey train (*osaru no densha*), 267, 276n7
Mukaiyachi Ikuyoshi, 34–40, 42, 45n10, 46n11, 47n15, 156

N

Naijo no kō (the wife's assistance), 152
Narrative *kyōzai* (disability memoirs used as educational material), 139
Nēmu (storyboarding), 119, 141, 252
Nichijō manga (slice-of-life manga), 86, 155, 175n10
Nigotta kanten (muddy agar), 202–203

Nihon Manga Kyōkai/the Japan Society for Studies in Cartoons and Comics (JSSCC), 280
Nihon Mangaka Kyōkai/the Japan Cartoonists Association (JCA), 92, 280
Niji-shōgai (secondary disabilities), 47n14, 124, 171
Ninchi no zure (cognitive gap), 137–139
Nippon Hōsō Kyōkai (NHK), 95, 121, 139, 169, 173, 174n6, 243
Nonami Tsuna, 11, 14, 124, 125, 145–173, 263, 283
Nothing About Us Without Us/*Nothing About Us Without Us*, 15, 28, 39, 42, 227

O
Obsessive-compulsive disorder (OCD), viii, 3, 11, 249–275, 284
Okita Bakka, 11, 14, 106n15, 113–142, 170, 174n8, 175n12, 225, 228, 283
One's safe zone (*jibun no kūkan*), 226
Onyu, 8, 127, 135, 165, 237

P
Page layout (komawari), 127
Panic attacks, 11, 41, 94, 95, 106n14, 183, 192, 260, 263–267
Peer support, 34, 41, 48n17, 61, 172
Personification (*gijinka*), 40, 41, 61, 193–199
Prime Minister Kishi Fumio, 280
[The] provoker/provoking (*tsukkomi*), 86, 99, 202, 228
Purification (*kegare wo harau*), 259, 260

R
Reading and OCD, 270
Rehabilitation (*shakai fukki*), 37, 137, 143n4, 171, 261
Rupo manga (reportage comics), 78, 82, 175n10
Ryōiku (special training for children with *hattatsu shōgai*), 118, 119, 142–143n4, 171
Ryōsai kenbo (the good wife and wise mother), 158

S
San-tōshin ("one-third is head"), 159
Scanlations (pirated digital manga translated by fans and amateur translators), 282
Seinen manga (mature men's comics), 155, 174n9
Seishin shōgaisha (mentally disabled individuals), 21, 22, 43, 51, 65
Selective mutism (*bamen kanmoku*), 117, 135–136
Selective serotonin reuptake inhibitor (SSRI), 203, 213n8, 231, 260
Self-help groups (*jijo gurūpu*), 61, 172
Self-transformation (*jiko hen-yō*), 59, 61, 234
Sengyō shufu ("professional house wives") versus *sengyō shufu* (full-time male homemaker) with a different kanji character for "fu", 158, 224
[The] Services and Supports for Persons with Disabilities Act (*shōgaisha jiritsu shienhō*), 223
Shachiku (company slave), 206
Shiramizu Sadako, 11, 14, 243, 249–275, 283
Shōgaisha (disabled individuals), 21, 24, 25, 143n7

Shōgai techō (disability passport), 135, 143n7, 170
Shōjo manga (manga for young female readers), 6, 78–84, 105n10, 105n11, 121, 129, 154, 156, 174n9, 223, 224, 241, 282
Shūdan seikatsu (group-oriented culture), 117
Sketching (*dessan*), 187
Sōgen (greenfield), 199
Suicide, 4, 27, 92, 122, 135–136, 203, 205, 212n3, 228, 231, 246n5, 285
  suicidal ideation, 8, 12, 95, 191, 231
Support for employment (*shūrō*), 59, 61
Sutōri manga (fictional story manga) as opposed to essay manga, 4, 6, 14, 77, 78, 83, 87, 90, 91, 251, 272, 276n8

T

Takiguchi Nozomi, 149, 174n7
Tanaka Keiichi, 11, 14, 94–96, 173, 181–212, 217, 221, 222, 228, 241
Tanki kioku shōgai (short-term memory disability), 124
Tankōbon (manga converted from magazine series into a book), 82, 147
Tantō henshūsha (manga editor), 113, 147, 184, 224, 252
Target readers, 152, 255
Tekiō shōgai (adjustment disability), 124
Tezuka Osamu, 93, 181, 186, 189, 190, 207, 213n5, 246n5
Tōbyōki (illness memoirs, illness narratives), 57–66, 69, 71n8, 95
Tōjisha, vii, 2, 3, 10, 11, 13–15, 21–43, 51–69, 78, 97–103, 106n13, 113, 115–118, 121, 123, 139, 140, 148, 150, 156, 170, 172, 174n6, 181–212, 217, 222, 223, 225–228, 242, 244, 245, 246n5, 251–256, 258, 261, 270, 272–274, 283
Tōjisha gatari (*tōjisha* storytelling), 34
Tōjisha honnin (the very person with a disability), 27, 106n13
Tōjisha kenkyū/self-directed research, 2, 13, 22, 26, 29, 33–43, 47n16, 47–48n17, 52, 60, 64, 66
Tōjisha manga, 2–4, 11–15, 22, 28, 42, 43, 56, 57, 59–61, 64–66, 69, 69n1, 71n9, 71n11, 77, 78, 80, 85, 91–97, 99–103, 106n15, 113, 116, 139, 193, 227, 228, 246n5, 252, 261, 282–285
Tōjisha mesen (*tōjisha's* perspectives), 22, 28, 29, 43, 94, 227, 251
Tōjisha narrative, 13, 14, 28, 51–69, 71n6, 95, 99–103, 139, 140
Tōjisha sei (tōjisha ness), 23, 24, 26, 27, 35, 59
Tōjisha shuken (the sovereignty of the tōjisha), 24, 29, 39
Tōjisha undō (movements by and for the tōjisha), 2, 22, 23, 29–31, 33, 38
Tomodaore (falling together), 226
Tsure (significant other), 217, 220, 225

Tsure ga utsu ni narimashite (*My SO Has Got Depression*), 8, 11, 14, 65, 173, 212n2, 217, 219
2011 earthquake-tsunami disaster, 253, 254

U
Uchi no OCD, 11, 14, 243, 249–275, 283
Uchū kaze (space cold), 241
US comics industry, 279, 281
"Useless bride" (*damena yome*), 157
Utsu-kun creatures, 199
Utsunuke: Utsu tonneru wo nuketa hitotachi (Depression Breakthrough: People Who Got through the Depression Tunnel), 14, 181–212, 217, 221, 228, 241
Utsu tonneru (the depression tunnel), 181, 183, 184, 188, 191–193, 199, 203, 207, 212n1
Uttōshi (depressing, dispiriting), 62, 227, 228, 242

W
Wing, Lorna Gladys, 150, 170, 171
[The] World Health Organization (WHO), 1, 10, 59, 276n1

Y
Yamai no katari (the talk of the illness), 61, 62
Yōkai (monster) versus *ūrei* (spirit), 192
Yon-koma manga (four-panel manga), 100, 120, 187
Yuru-kyara character, 100

GPSR Compliance
The European Union's (EU) General Product Safety Regulation (GPSR) is a set of rules that requires consumer products to be safe and our obligations to ensure this.

If you have any concerns about our products, you can contact us on

ProductSafety@springernature.com

In case Publisher is established outside the EU, the EU authorized representative is:

Springer Nature Customer Service Center GmbH
Europaplatz 3
69115 Heidelberg, Germany

www.ingramcontent.com/pod-product-compliance
Lightning Source LLC
Chambersburg PA
CBHW071921011225
36163CB00005B/319
* 9 7 8 3 0 3 1 0 0 8 3 9 9 *